Rabbits

by Chris McLaughlin

An Imprint of BowTie Press®
A Division of BowTie, Inc.
Irvine, California

Lead Editor: Jennifer Calvert
Art Director: Cindy Kassebaum
Production Supervisor: Jessica Jaensch
Assistant Production Manager: Tracy Vogtman
Book Project Specialist: Karen Julian
Indexer: Melody Englund

Vice President, Chief Content Officer: June Kikuchi
Vice President, Kennel Club Books: Andrew DePrisco
BowTie Press: Jennifer Calvert, Amy Deputato, Lindsay Hanks, Karen Julian, Jarelle S. Stein

Front Cover Photography: (main) Fábio, do Blog Peladeiros (left inset) Rosalia Wilhelm (right inset) Maja Dumat
Back Cover Photography: Kassia Halteman/Shutterstock

Library of Congress Cataloging-in-Publication Data

McLaughlin, Chris.
Rabbits : small-scale rabbit keeping / by Chris McLaughlin.
p. cm. -- (Hobby farms)
Includes index.
ISBN 978-1-933958-96-5
1. Rabbits--United States. I. Title. II. Title: Small-scale rabbit keeping. III. Series: Hobby farms.
SF453.4.M35 2012
599.32--dc23

2011031192

BowTie Press®
A Division of BowTie, Inc.
3 Burroughs
Irvine, California 92618

Printed and bound in China
15 14 13 12 11 1 2 3 4 5 6 7 8 9 10

This book is dedicated to Leo and Laughing Crow—
two special rabbits that just rocked our world.
And to all the animals who have crossed paths
with me and have been so patient while I learned:
you have been my greatest teachers.

Also to rabbit man extraordinaire Bob Whitman,
whose love of rabbits and dedication to the rabbit
world was an inspiration to all of us.

CONTENTS

INTRODUCTION

Why Rabbits?

Time and time again, rabbits have proven themselves an ideal livestock choice for small farms. In a world where farming is synonymous with land and acreage, hobby farms create an opportunity to live your rural dream in a smaller space. Rabbits are the perfect hobby-farm livestock, as they require very little space. Even hobby farms located in suburban and urban neighborhoods will find small-scale rabbit keeping an easy endeavor.

Raising rabbits demands little in the way of financial resources or specialized equipment, and the rewards are considerable. These petite creatures can offer companionship and the chance to participate in competitions. On a more practical level, they can provide meat and fiber (wool). Another valuable by-product of this small livestock is top-of-the-line manure for your garden and compost piles. Rabbit manure is widely recognized by gardeners as the most nutritionally balanced manure of all the herbivores. (It's also an excellent medium for raising fishing worms!) And you don't have to choose just one reason to raise rabbits. For example, a rabbit raised for fiber can also be a beloved family pet.

We brought home our first rabbit seventeen years ago, and we still consider rabbits a perfect fit for small-scale hobbyists. Rabbits are easy to house, handle, care for, and transport. They're also curious, affectionate creatures that are extremely entertaining. Turn the page to see why so many small-scale farmers are touting rabbits as one of the most popular little livestock choices around!

chapter **ONE**

Meet the Domestic Rabbit

As you may have guessed, rabbits didn't start out as pets. In fact, the term *domestic* in the general sense of the word doesn't refer to being a pet at all. When animals are domesticated, they're kept and cared for by people for any number of purposes, be it for fur, meat, fancy (show), or—yes—companionship as family pets. The domestic rabbits we see in the United States today are a direct result of European settlers' shipping Belgian Hares (they were actually rabbits, not hares) to America in the early 1900s for breeders and collectors. Later, in 1913, New Zealand Red rabbits were brought to America by sailors who kept them on board as small and easy-to-care-for meat animals. Since that time, many more rabbit breeds have been imported from countries worldwide. And the rest, as they say, is history.

Rabbit Ranking

In the animal kingdom, the class Mammalia holds a vast number of orders and suborders. Animals are classified according to their traits and habits to make it easier to observe and study them. Rabbits belong to the order Lagomorpha. This name comes from the Latin words *lagos*, meaning "hare," and *morphe*, meaning "form." The order has only two families: Leporidae (rabbits and hares) and Ochotonidae (pikas).

Rabbit History 101

According to geologists, rabbits have been hopping about the planet for 30 to 40 million years. It's said that the Phoenicians kept caged rabbits as far back as 3,000 years ago, and the Romans followed suit shortly thereafter. Eventually, people in southwestern Europe, on land that's now considered part of Spain and Portugal, saw the wisdom in *raising* rabbits for meat (instead of hunting them), and so began the heritage of the domestic rabbit. Of course, these rabbits weren't anything close to resembling pets. Rather, they were an extremely convenient, easily transported meat source.

Hare's Looking at You, Kid

Rabbits, with their curious, kind, and fast-footed habits, have captured the imagination to the point where we've placed them in the center of mainstream media by way of folklore, literature, advertisements, and entertainment. Check out some of the most famous rabbits in the world:

- **Bugs Bunny—**With his world-renowned and charming phrase, "What's Up, Doc?," Bugs Bunny is perhaps the most famous rabbit anywhere. He was created in 1940 by Looney Tunes and Merrie Melodies (which, in 1944, became Warner Bros.).
- **Harvey**—This was James Stewart's 6-foot, 3½-inch tall, invisible rabbit friend in the 1950 movie *Harvey.*
- **Energizer Bunny**—The pink bunny with a drum that just keeps going, and going, and going showed up as the spokesbunny for Energizer batteries in 1989.
- **Roger Rabbit**—He was Gary Wolf's main character from the 1981 novel *Who Censored Roger Rabbit?,* and he was later transformed into a cartoon character in the 1988 movie *Who Framed Roger Rabbit?*
- **Peter Rabbit—**Peter is the adventurous bunny in Beatrix Potter's children's book *The Tale of Peter Rabbit,* which was first privately printed in 1901.
- **Peter Cottontail**—This Peter lived in the 1914 children's book *The Adventures of Peter Cottontail* by Thornton Burgess.
- **The White Rabbit**—Alice follows the white rabbit down the hole in Lewis Carroll's (Charles Lutwidge Dodgson) 1865 novel *Alice's Adventures in Wonderland.* In this same book, Alice visits with the **March Hare** and the Mad Hatter.
- **Velveteen Rabbit**—In Margery Williams' 1922 children's book *The Velveteen Rabbit,* this is the toy rabbit that becomes real thanks to the love of a child.

- **Br'er (Brer) Rabbit**—This tricky rabbit is linked to both African and Cherokee cultures but was adapted by Walt Disney in the animated feature *Song of the South.*
- **The Trix Rabbit**—Remembered by baby boomers for the tagline of "Silly rabbit, Trix are for kids!," the General Mills cereal-box mascot was created in 1959 by Joe Harris.
- **Thumper**—He was a deer's best friend in the 1942 Walt Disney classic animated movie *Bambi.*
- **"Rabbit"**— He is the gardening enthusiast from the children's stories of *Winnie the Pooh,* written by A.A. Milne in 1926.
- **Bunnicula**—He is the vegetable-juice-sucking character from the 1979 children's book written by Deborah and James Howe, *Bunnicula.*
- **Uncle Wiggly**—He starred in the original 1910 children's book *Uncle Wiggly* by Howard Roger Garis.
- **Hazel, Fiver, and Blackberry**—The entire cast of 1978 animated film *Watership Down* was made up of rabbits.
- **Leo**—He is the main character in Stephen Cosgrove's 1977 children's book *Leo the Lop,* part of the Serendipity Series.
- **The Hare**—This is the guy who took too much for granted as he snoozed and lost the race to the tortoise in Aesop's fable *The Tortoise and the Hare.*

As people migrated from land to land by ship, rabbits either escaped at port or were released from their cages. These rabbits made themselves at home wherever they landed, which created wild populations of *Oryctolagus cuniculus* throughout Europe. Around the sixth century AD, French Catholic monks began raising rabbits for meat. Keeping rabbits within the monastery walls led to breeding them for various uses, sizes, and colors. These monks are still credited today with the first true domestication of the rabbit in Europe.

Although European settlers introduced the Belgian Hare to America in the early 1900s, we know that Lop and Angora rabbits were already prevalent in the United States by the mid 1800s. In 1910, thirteen people formed the new National Pet Stock Association of America—an organization that would later become the American Rabbit Breeders Association (ARBA), which now boasts an estimated 30,000 members and recognizes nearly fifty different rabbit breeds. Today, rabbits are fourth only to dogs, cats, and birds as pets in the American household.

The wild rabbits you see, like this one, are descendents from European rabbits.

The Rabbit Today

Depending on where you make your home, rabbits may be kept as companions or family pets and live indoors or outdoors in an enclosure. In fact, some people become so closely bonded to their furry friend that their rabbit is given its own bedroom in the house! Other rabbit fanciers raise their rabbits in a "rabbitry" (a rabbit-specific area or enclosure where rabbits are raised and cared for) as a hobby for showing. Angora rabbit breeds are raised for both show and fiber (fur) because they produce wool that can be spun for knitting. Rabbits are small, easy to care for, and provide healthy meat, which also makes them a smart choice for breeding and raising for food. The savvy rabbit-raiser will research the individual rabbit breeds and choose one that catches his or her interest as well as incorporates one or two of the uses mentioned above. With their comparatively small size, gentleness, and versatility, it's clear why rabbits are one of the most popular hobby-farm animals today.

Splitting Hares

Rabbit species are often confused with one another. These furry, hopping creatures all look very similar, but our domestic rabbit species isn't nearly the same as the American wild rabbits (called *cottontails* or *brush rabbits*). In fact, they're cousins, at best.

The rabbits that we call pets and companions here in the United States have all been domesticated from the European rabbit, *Oryctolagus cuniculus.* These European rabbits live in family colonies and create an underground system of warrens or burrows, which they use to hide from predators and to give birth. Cottontails are of the genus *Sylvilagus* and lead a different lifestyle than that of their cousins. They live above ground and give birth above ground in depressions in the earth. Cottontails will, however, use burrows that were created by

The hare may be your rabbits' cousin, but they're easy enough to tell apart.

Mousy Tendencies

Rabbits, hares, and pikas (all from the order Lagomorpha) have something in common with rodents: their teeth continue to grow all of their lives. This makes it necessary for both rodents and lagomorphs to chew constantly to keep their teeth worn down.

other animals or brush piles as temporary protection against predators.

To set the record straight, both the jackrabbit and snowshoe rabbit are actually *hares*, while the Belgian Hare is really a breed of *rabbit*. Like rabbits, hares (of the genus *Lepus*) also belong to the family Leporidae, but there's quite a bit of difference between the two. One of the major differences relates to their young. Baby hares are called *leverets* and are born with more survival advantages than rabbits. Newborn leverets are *precocial*, which means they come into the world completely furred and with their eyes open. They can hop around about an hour later and begin nibbling the grass around them within a week. Rabbit *kits* are born the polar opposite: they're *altricial*, which means they're completely hairless and blind at birth, obviously requiring more care.

A mother hare spreads her litters out in several different depressions on top of the ground. This tactic leaves the little leverets (baby hares) more exposed than the babies of tunnel-digging rabbits. But because the little leverets are capable of seeing and moving around when they're born, they're able to hop about, hide, and return to their birthing area when the mother hare is around to nurse them.

Several other characteristics differentiate hares from rabbits: Hares have longer back legs than rabbits, making them stronger and faster. Their ear length is also exaggerated when compared to rabbits' ears. Unlike rabbits that live in colonies in the wild, hares live a solitary life; they pair up only for mating and then go their separate ways. In addition, hares rely on their speed to escape predators, while rabbits tend to hide in burrows or brush piles. Finally, European rabbits have been domesticated as pets; this has never been the case with hares.

Domestic Rabbits in the Wild

If your domesticated rabbit escaped its hutch, it would instinctually dig a burrow. But that's as far as its survival skills would get it. The sharp instincts necessary for a rabbit to survive in the wild have been watered down by generations of domestication. This is why seasoned owners cringe when someone happily reports that he or she has "set a rabbit free into the wild." Though the person feels he or she has done the right thing, the freed rabbit won't survive for very long.

One thing working against domesticated rabbits in the wild is coat color. Fanciers have bred many colors in rabbits for people to enjoy, but these unnatural colors do not necessarily blend in with natural surroundings. This makes the domestic rabbit a beacon to every predator in the area, including hawks, foxes, coyotes, raccoons, and even domestic dogs. Of course, there are domestic rabbits with the *agouti* coat color, a grizzled brown closest

Domestic rabbits need supervision in the great outdoors for many reasons. One is that they stand out from their surroundings, making them a prime target for predators.

to the color of their wild ancestors. Rabbits of this color may have a small advantage over their unnaturally colored brethren, but there are other essential survival instincts that a domesticated rabbit simply does not have, such as the ability to detect or escape predators. Domestic rabbits are also heavier than wild rabbits, which means that they can't run as fast.

Unfortunately, while a pet rabbit may hop away and try to hide, it just isn't equipped to survive on its own for long. Even a wild cottontail rabbit has a lifespan of only about a year—possibly three, if it's very, very clever. If a freed domestic rabbit survives as long, it's due to sheer luck and nothing else. The moral of the story? Never let a domestic rabbit loose in the wild!

ADVICE FROM THE FARM

The Rabbit Habit

Our experts share some general rabbit-raising information.

Rabbits Just Make Sense

"With current trends in sustainable suburban agriculture, rabbits just make sense as the perfect new 'livestock' for virtually anyone!"

—Allen Mesick, ARBA Judge

Start Small and with Help

"Friends of ours in the 4-H group, who had been participating in the group for many years, suggested we try the 'rabbit project.' What started as a second-grader's first-year project for 4-H has evolved into a rabbitry with over twenty Netherland Dwarfs showed nationally through various rabbit breeding groups. As the mother of the rabbit breeder, I would highly recommend the 4-H rabbit-breeding project to any parent looking for an educational program to teach children basic animal-care skills and responsibility."

—Brenda Haas

No Stupid Questions

"I would like to stress how important it is to have the courage to ask questions when in doubt. Everyone started their rabbit-raising hobby somewhere and got to where they are today because of the help others gave them. Most rabbit-raisers are willing to share their tips and tricks for successful raising if you ask. We raise rabbits for a hobby. The competitive drive [we exhibit], friendships we make, animals we produce, shows we attend, and crazy things we do for our animals are all done out of passion for the hobby. It is important to enjoy the hobby, learn as much as you can, and have fun!"

—Keelyn Hanlon

A Little Perspective

"To see the world from a rabbits' perspective, get down on the ground and look up at people walking around. The world looks a lot different from this height."

—Cassandra Brustkern

chapter **TWO**

The Versatile Rabbit

Because of their soft coats, small stature, and gentle nature, rabbits have the reputation of being terrific pets. But rabbit enthusiasts know that this furry creature has many other benefits, especially as livestock on a hobby farm. These versatile animals can be raised for fiber (wool), meat, and manure as well as for showing in the 4-H and competition circuit. The beauty of this animal is that a single rabbit breed can perform many of these duties at once. Use the information provided in this chapter to help you decide what you want from your rabbits. Then, once you've got your rabbitry up and running, you can return to this section for instruction on how to do what you've chosen to do.

The Right Rabbit

With nearly fifty rabbit breeds recognized by the ARBA (visit www.arba.net/breeds.htm for a list) and a vast array of coat colors and patterns, how do you decide which breed is best for your rabbitry? Well, start by determining what you'd like to do with your future livestock. Do you want to make money by selling them for meat, fiber, or fancy? Do you want to show your rabbits? Do you have children that are looking for a new hobby and can show them at local fairs? Are you a knitter or spinner who's tired of paying high prices for valuable rabbit wool and wants to make a little cash by selling your surplus fiber?

You'll save yourself a lot of time and trouble if you decide what you want at the outset instead of purchasing rabbits on a whim and slowly migrating toward what you'd really like to do with them. And don't forget that many rabbit breeds can serve multiple purposes, so choosing a dual-purpose breed makes sense for many rabbit keepers.

The next step is to look over the lists of breeds in the Resources section to see if any attract you. Go to the ARBA website (www.arba.net) and look at all of the recognized breeds, then make contact with a club or breeder that specializes in the breed that catches your eye. These people can give you a better understanding of their breed and help you decide whether it's the right choice for you and your plans.

Whether you're planning on showing rabbits or not, I highly recommend attending a large rabbit show and perusing the aisles for a few hours. Even if not every breed is there, many of them will be. Ask the owners about their rabbits, ask to pet one, and ask why they chose that breed. Stand around the judges' tables and listen to what they're looking for in a winning animal. Most judges are more than happy to answer questions either during a break time or after the judging. (None are willing to answer questions in the middle of a class.)

Rabbits as Pets

Rabbits have kept humans company as long as any domestic animal—and for good reason. They're small, gentle, soft, and extremely quiet. They're easy to keep and they have very few specific needs.

First Things First

Before you jump into creating a rabbitry and purchasing animals, ask yourself some important questions. Here are a few to get you started. Take the time to answer them honestly and thoroughly to clarify your needs.

- How much space do you have? Enough for a few rabbits or an entire rabbitry? Ideally, you want your rabbitry to be on the coolest side of the property. Part of your answer will also be based on the size of the rabbit breed that you choose. For instance, the amount of cage space you'll need for five French Lops could house ten Netherland Dwarfs. Don't forget to add some space if you'd like a play yard where the rabbits can be turned out to run, dig, or nibble on grass.
- Can you provide protection for your rabbits (such as an overhang or enclosure) against weather and predators?
- Will your rabbitry bother the neighbors? A large rabbitry could emit odors. If you think this may be an issue, do you have space in a more isolated area?
- Does your city allow rabbit keeping? Many cities and homeowners associations do not allow livestock keeping in their communities. If they do, some may require a license or permit for the hobby. Be sure to research the requirements of your community before buying any animals.
- How much help do you have for rabbitry maintenance? If you're relying on family members for help, ask yourself which tasks they're truly capable of performing.
- Which facets of raising rabbits appeal to you? Companionship? Wool or meat? Manure to enrich your garden soil?

Rabbits make great pets, especially if you take the time to handle and socialize them.

Pet rabbits don't require vaccinations and can be kept in even the smallest apartments.

Rabbit owners that take the time to interact with and watch their furry friends find that rabbits have individual personalities and can be very entertaining. (The juveniles are the most fun to watch with their air-springing antics!) That said, rabbits are easily startled and can scratch young children. Rabbits' legs and backs are also easily injured, making it important to supervise kids when they're handling them or caring for them.

Although rabbits prefer to have all four paws on the ground, once a rabbit is completely comfortable with you, it'll relax its natural instinct and lie on its back in your arms. Now that's trust! Most rabbits are up for some lap time that involves petting and chatting. The best way to do this is to place a towel on your lap to prevent scratches from your bunny's toenails should it try to hop away.

Rabbits can become very fond of their caretakers, but this can only be achieved through consistent contact with them. Playtime also keeps rabbits physically and mentally stimulated.

Raise a Friendly Rabbit

Always remain aware that, first and foremost, your rabbit is a prey animal (see chapter 4 for more information about rabbit behavior). As rabbit-raisers, the only way to avoid evoking rabbits' defensive behaviors is to gain their trust. But humans are predatory animals, and rabbits instinctively know this. Thankfully, hundreds of years of domestication are on your side. Rabbits bred to be pets are distant cousins to the wild rabbits in the world today, so it shouldn't take long for you two to become friends.

Common Pet Rabbit Breeds

It's worth mentioning that any rabbit breed can make a great companion. It truly depends upon the individual rabbit's personality. That said, the majority of pet rabbits are in the lightweight and small categories. Popular rabbit breeds kept as pets include:

- American Fuzzy Lop (4 pounds)
- Dutch (5½ pounds)
- Dwarf Hotot (2–3 pounds)
- Himalayan (4½ pounds)
- Holland Lop (4 pounds)
- Jersey Wooly (3 pounds)
- Lionhead* (2–4 pounds)
- Mini Lop (6½ pounds)
- Mini Rex (4½ pounds)
- Mini Satin (4 pounds)
- Netherland Dwarf (2 pounds)
- Polish (2–3 pounds)

*Lionheads are not recognized by the American Rabbit Breeders Association as a true breed yet. However, they're gaining popularity as pets.

Teach children how to properly handle a rabbit so that everyone involved is happy.

If the rabbit is meant to be a pet for an adult, all breeds can and should be considered. But choosing a pet rabbit for a child creates some extra concerns. It's extremely important to think about the size of the child. An 8-year-old may find it pretty difficult to pick up and handle a 17-pound French Lop, for example. It's not impossible; it just may not be optimal. Angora breeds also don't make great pets for kids. Although Angora rabbits are extremely friendly animals, and thus suited temperamentally to children, they require too much grooming work for the average youngster to handle (so, if you get one for your child, expect to pitch in quite a bit!). Also keep in mind that rabbits such as American Fuzzy Lops and Jersey Woolies are considered to have wool as opposed to normal fur, but both breeds grow considerably less than their full-wool counterparts, such as the English Angora. In addition, after their baby coats fall out, a different adult coat comes in. This adult coat has a lot more of what are referred to as "guard hairs," which makes the coat far less wooly and more manageable. It's important to research the different breeds and their care rather than simply rely on their names, because as you can see, first impressions can be deceiving.

Rabbits in the Spotlight

There are several ways to have fun with your rabbits that have nothing to do with raising them for their by-products. Showing rabbits is a hobby

Handle with Care

Here are some techniques that will help turn your rabbit into your best friend.

- Handle your bunny for short periods *every day*. It isn't the length of time that you hold and pet it that's important; it's the frequency.
- Start out in a small, confined space with your rabbit so that if it hops away, you won't be forced to chase and catch it.
- Talk to your rabbit and approach it calmly. Loud, sharp noises will provoke its flight response.
- Try letting your rabbit out of its cage (in a bunny-safe environment) and let it come to you. Rabbits are curious creatures; you may be surprised at how quickly it approaches you.
- In the beginning, simply sit and pet your rabbit (as opposed to holding and carrying it). A rabbit that's being carried has lost its greatest defense—the ability to run away from danger. Give your bunny some time to figure out that you're not a threat.
- Offer healthy treats or rabbit-safe toys when you play with your rabbit. Let it associate you with all things tasty and fun.
- Explain to visitors that they need to handle your rabbit the same way you do. If your friend grabs or chases your bunny, it could set your relationship with your rabbit back for weeks.
- Don't let predators such as dogs or cats near your new pet; their natural instinct is to hunt rabbits. Besides terrifying your rabbit, they could also cause serious injuries.
- Keep your rabbit's home clean and provide fresh bedding, food, and water at all times. A healthy bunny is a happy (and therefore friendly) bunny.

in which human and animal bonding is encouraged and rewarded with titles, ribbons, and other prizes.

Raising Show Rabbits

Raising show rabbits is often referred to as raising rabbits for *fancy*. If your small livestock is papered (having documentation proving its purebred lineage), and it's a breed that's recognized by the ARBA, then you can become involved in the rabbit fancy. Only purebred animals are shown.

Rabbit shows quickly take on a life of their own as a rabbit breeder becomes part of the show community. Showing rabbits is a fun and entertaining hobby that can end up becoming a small business venture, as well. Many a rabbit fancier has started showing his or her animals "just for fun" only to become hooked for life.

One of the perks of showing rabbits, as opposed to dogs or horses, is the expense (or lack thereof). Rabbit shows are quite inexpensive to participate in, so

Flemish Giant Samantha spends her days enjoying the spotlight and the spoils.

they appeal to the fancier who loves the idea of some healthy competition and who takes pride in winning but doesn't want to blow his or her family's budget. Other allures of joining the show circuit include the ability to breed a higher quality of rabbit or simply socializing within a community of like-minded people. Many breeders at the shows use their stock for purposes other than (or in addition to) showing, and it's always nice to cultivate a wider perspective when raising livestock; there's a wealth of knowledge to be gleaned from the rabbit judges and the seasoned rabbit breeders and exhibitors at these competitions.

Participating in a rabbit show can be exciting, exhilarating...but often confusing.

A Word to the Wise

Before you purchase rabbits for the show table, check out the American Rabbit Breeders Association website (www.ARBA.org), and contact the national specialty clubs for any breeds that strike your fancy. These groups can put you in contact with local breed clubs and breeders—all of whom you can trust to be legitimate if they're affiliated with the ARBA. Like everything else in life, there are people who are more than willing to take advantage of a newcomer. Plus, it just makes good sense to become very familiar with the breed you are interested in showing before you take the plunge.

Show Rabbits Sport Tattoos

Rabbits that are entered in shows are tattooed in the left ear using permanent ink. The tattoo can have numbers, letters, or a combination of both, and often it gives an indication of the rabbitry where the rabbit was born. The ear tattoo is how a rabbit is identified at shows; it also appears on the rabbit's papers.

Many owners tattoo their own rabbits using purchased kits. Here's how it works: An ear clamp holds pins that make up the tattoo letters and numbers. You clamp the rabbit's ear, causing the pins to pierce the tissue of the ear, and then you rub ink into the resulting piercings. The tissue heals over the ink, creating a permanent mark. You can also use a tattoo pen to write the letters or numbers inside the ear. Tattoo pens are popular with a lot of breeders because you can write the letters or numbers out in freehand as opposed to using the tattoo clamp, which presses a bunch of needles into the ear at once. Many people swear by the tattoo pen because it's pain-free and stays in the ear longer without having to be redone.

In either case, the rabbit needs to be held firmly by one person while another does the tattooing. Some technique is required to avoid the large vein that runs through the rabbit's ear. Also, if the task isn't handled well, the rabbit can struggle and flip off of the table or out of someone's arms. Many a rabbit has broken its back due to improper restraint while being tattooed. For this reason, you should learn how to tattoo—hands-on—from an experienced rabbit person before you try it on your own.

On occasion, a tattoo will fade to the point of being unreadable. Unfortunately, you may notice this when you're already at the show; because of this, you should always travel with your tattoo pen. Be sure to bring a rag and something to clean the ear with, as well.

There's a lot going on and it's all happening at different show tables at the same time. To the novice rabbit owner, it looks chaotic—people rushing to retrieve rabbits from their carrying cages, exhibitors grooming, breeders purchasing, ribbons flying—but I assure you, there's a method to the madness.

To get the most out of your first show experience, be sure to tag along with a seasoned show person. If you don't know any, contact a local rabbit breeder to see if he or she will be attending the show. Ask if you can meet at the show to ask a few questions. Rabbit people are usually very friendly and willing to help—mostly because they love talking rabbit! Once you arrive, check in at the registrar's table, and ask any questions you may have. Find out where and when your rabbit breed is showing. You don't want to miss seeing your breed in action!

If you're visiting a show without actually showing, that's an even better way to learn. You can still ask the registrar or breeders questions, and you'll be much more relaxed about it (considering you don't have to listen for your breed to be called over the loudspeaker).

Pedigree Papers

You may have considered showing rabbits but have yet to step foot inside a rabbit show. Or maybe you've been to a few shows and have caught a touch of rabbit fever. There's also the possibility that, until this section, you had no idea that rabbit shows existed at all and are giggling at the thought. (Well, there really are rabbit shows, and the people at the top take it very seriously, indeed.) Perhaps you have some purebreds and just want to learn more about them. In any case, as long as your rabbits have pedigrees, or "papers," you can join right in!

Only purebred rabbits are entered in shows because each breed has its own standard. This is a set of guidelines by which a purebred animal is judged. The idea is for each animal to be the best representation of its breed. The American Rabbit Breeders Association has a book available for purchase called *The Standard of Perfection* that includes the standards of every rabbit breed recognized by the ARBA.

A written pedigree should accompany any purebred rabbit that you purchase; it should include four to five generations of that rabbit's heritage. These papers show the accomplishments of the parentage of your rabbit on both sides of its family tree. This is considered proof that the rabbit is the breed the breeder claims it to be, and it shows the weights and winnings of its family members. This helps people make wise choices when choosing stock for breeding.

Papers aren't necessary if you don't plan to show your rabbit. But if you'd like to join the show circuit, sooner or later you'll need them. The best idea is simply to purchase purebred stock with papers, no matter what your intentions are. Having papers gives you flexibility with your choices; you can always try a rabbit show later if you want to give it a go. It also makes it much easier to resell your stock should you want to try a different rabbit breed down the road.

Registration Papers

A rabbit that has "papers" is simply a rabbit that has a pedigree. Registration papers, often confused with pedigree papers, are a different thing altogether. There's no urgent need to register your rabbits. In fact, it's not required for show purposes. The advantage of having some or all of your animals registered is that

Get Ready to Show

It seems like no matter how hard we try, something essential still gets left behind! Here's a handy checklist to help you be sure you have everything you need for a one-day rabbit show.

What to Pack for You:

- Activities for your young children
- ARBA membership card and the original pedigrees for each of your rabbits if you plan on registering them. Many people are concerned about bringing original pedigrees to a show, but you can always leave copies at home.
- Business cards
- Camera with fresh batteries
- Cell phone, fully charged
- Directions to the show
- Folding chair(s)
- Glasses, reading and sun
- Pens, scissors, markers, tape (honestly, you never know)
- Personal care items: bandages, lip balm, eye drops, pain reliever, hand sanitizer, antihistamines, any necessary medications
- Show coat or apron for handling and grooming rabbits
- Small cooler with snacks and beverages
- Sweatshirt or jacket

What to Pack for the Rabbits You're Showing:

- A copy of the ARBA's *The Standard of Perfection* (to clarify possible disqualification from a judge or to use as a reference when purchasing another rabbit)
- Bottle of water (some rabbits won't drink unfamiliar water)
- Cart or wagon to transport carrier(s) in and out of the building
- Entry form to confirm the rabbits were entered in the show
- Extra rabbit carriers if you plan on purchasing rabbits
- Grooming supplies
- Hay and rabbit pellets
- Tattoo kit for touch-ups

Finally, don't forget to bring the right rabbits! Just before you leave, double-check your entry form for ear numbers and match them with the rabbits you've loaded up.

they have been looked at and approved by a registrar who is licensed by the ARBA.

What exactly are they approving? Well, any rabbit that is purebred can have a pedigree, but that doesn't necessarily mean that the rabbit is a good example of that breed. For instance, even a purebred, pedigreed animal can be over the breed standard's weight limit, have a wrong-colored toenail, have a malocclusion (misalignment of the upper and lower jaw), or display bad overall coloring—all genetic characteristics that could disqualify the rabbit from the show table.

A registrar goes over each animal individually and makes certain that the rabbit is a good representation of the breed. He or she will only register those rabbits that fall within the breed's standard and that have no disqualifying characteristics. If the rabbit meets the registration qualifications, he's tattooed with a registration number in its right ear (the left ear holds the show tattoo), and registration papers are sent to the ARBA by the registrar. In a few weeks, you'll receive a certification paper for the rabbit.

A registered animal proves to potential buyers that you're producing quality animals in your rabbitry—that your rabbits are excellent representations of the breed. This isn't a guarantee that they'll win on the show table, but this added step will allow you to sell your rabbits for more money.

Registrars are often at the larger rabbit shows, and some will come to your home if you have a fair number of animals to register. Often, breeders will get together and have their animals registered together to make it worth the registrar's time to make the trip out. In order for a rabbit to be registered, you need to have with you:

- The registration fee
- Proof that you're a member of the ARBA

Your rabbits will need to be even-tempered to shine on the judges' table.

- A copy of the rabbit's pedigree
- The rabbit (obviously)—with a permanent show tattoo in its left ear

Where the Rabbit Shows Are

Now that I have you interested in attending a show, how do you find one? Unfortunately, rabbit shows aren't widely advertised in the local papers. Seek out specific breed clubs or state clubs, which have all the information you need on local shows. There's a list of national specialty clubs in the Resources section of this book, but you can also contact the American Rabbit Breeders Association (ARBA) or use its website to find clubs near you.

You don't have to join a club to get information on the rabbit shows that are held in your area (or the areas closest to you). Ideally, what you want is the name and contact information of one of the show secretaries—get on her mailing (or e-mailing) list for all the shows for the year. Rabbit exhibitors at your local fair will often have information for you, as well, because many of them also show outside of the fair.

4-H clubs are a great way to connect with your community and the animals you love.

4-H Rabbits

One of the best ways to enjoy your family pet is by joining a youth organization called 4-H. (The four Hs stand for *head*, *hands*, *heart*, and *health*.) You can get in touch with the 4-H groups in your area by contacting the cooperative extension office in your county. The 4-H club has a variety of individual projects that run the course of the school year and end with the kids entering the items they've made, the produce they've grown, or the animals they've raised in county fairs. Not only is there a rabbit show at the fair, but the 4-H kids also have many different rabbit breeds on exhibition for the public. Because 4-H kids are so hands-on with their animals, these rabbits are usually the friendliest around.

In the 4-H rabbit project, members are expected to learn how to care for their animals properly and are required to do so for credit in the club. They learn about rabbit husbandry, health, first aid, conformation, breeding, and handling. 4-H also teaches general responsibility, record keeping, and expense tracking. Kids also learn how to present what they've learned at public gatherings such as in the classroom, at the fair, or at community meetings.

When it comes to showing rabbits at local fairground youth shows, the animals don't need to have papers. The rabbits do need to be purebred, but they don't need the pedigree to prove it (it's too easy for kids to lose the papers, so the judges don't require them unless they believe the rabbit may be mixed). If you see a rabbit breed that you're interested in at a fair, ask a 4-H student at the small animal exhibit about it. You'll be impressed by the member's knowledge.

Rabbit Hopping (Jumping)

Rabbit hopping or jumping is an exciting sport that originated in Sweden in the 1970s. While it has yet to catch on in the United States the way that it has in Denmark, Norway, and Germany, it's a sport that may just attract your interest. Rabbit hopping is worth considering for young rabbit owners and 4-H club members because it fosters a close bond between children and their pets. Rabbits don't have to be papered or purebred to participate, which opens the hobby up to kids with crossbreeds and rescued rabbits. The only requirement is that the rabbits need to be at least four months old.

At these events, rabbits have been previously trained to wear both a harness and leash. They're then taught to jump over small jumps, such as short, decorative little fences, which are laid out in patterns similar to those in horse show-jumping competitions. As in traditional rabbit shows, these hoppers earn points for each jump that eventually lead to championship titles. A rabbit's score (and ultimately, its win) is based on how high and how wide an obstacle the rabbit can jump as well as how well it clears them without *faults* (knocking down a fence rail). Hopping is also a timed event.

Jumping is wonderful physical exercise for rabbits and is mentally stimulating for them, as well. Beginning hoppers are put on a straight course with jumps that are 4, 6, 8, and 10 inches tall, but advanced rabbits jump much higher. These miniature jumps are just about as creative and colorful as the builder wants them to be, and because they're handmade,

No, this course isn't for very small horses—this colorful scene is a rabbit hopping show.

Rabbit hopping and agility let rabbits showcase their natural abilities.

dads and other building-inclined family members are naturally encouraged to join in the fun.

Rabbit-hopping rules were developed primarily for the safety and well-being of the animal; the rules exist so that the sport never becomes cruel or stressful for the rabbit. Check the Resources section for more rabbit-hopping information, or contact your local 4-H office. If you can't locate a club near you, consider starting one yourself!

Another sport, rabbit agility, looks very similar to rabbit hopping, but there are some differences. With hopping, the rabbits are on leashes, and owners run alongside the rabbits as they jump the obstacles. The only equipment within the course is the jumps. With rabbit agility, the rabbits are off leash, and the course may or may not be fenced (this is up to whoever is running the particular event). The agility course is made up of not only jumps but also other agility equipment, such as chutes and holes, that must be maneuvered, as well. Rabbits participating in agility learn how to go over, under, around, and through the various pieces of equipment. Like hopping, agility is also timed.

Raising Rabbits for Fiber (Wool)

Angora rabbits are well loved for their kind temperament and friendly attitude. But they're best known and sought after for their plush, silky, cloudlike coats. Angora wool is in high demand and considered top-drawer in the fiber market. It's among the softest of garment fibers in the world.

If raised primarily as wool-producing animals, Angora rabbits are considered *no-kill livestock,* which is very appealing to a lot of would-be rabbit farmers. Rabbit fanciers that raise Angoras can feel free to get emotionally attached to their livestock. (In addition to fiber, commercial breeders may also raise Angoras for showing and meat, tripling their investment.)

The Angora rabbit is thought to have originated in Ankara, Turkey, although the hard facts remain unclear. What is certain is that breeders in Europe have raised Angora rabbits for their fiber for centuries. The French get the credit for making Angora wool popular around 1790; North America didn't see Angoras until 1920.

About Wool

Angora fiber, which is seven times warmer than sheep's wool, creates an extremely cozy fabric. It can be sold either spun or raw and can be dyed or left its natural color. Angora wool is so fine that some spinners prefer a fiber blend to hold dense-knit stitches. The wool is often blended with other fibers such as sheep's wool, mohair, silk, and cashmere.

Rabbit wool is harvested by either plucking or shearing. Most Angora rabbits will naturally shed their coat three to four times a year. Breeders

If you're interested in spinning your own yarn, either for personal use or to sell, a rabbit is certainly more economical to keep than a sheep.

can take advantage of this by plucking during this time. In many parts of the country, modern show rabbits are bred to hold on to their long coats; these types of Angoras may need to be shorn rather than plucked.

A breeder can also shear the wool off of the rabbit using scissors or clippers every ten to eleven weeks, depending on the desired length of the wool. In the case of the Giant Angora, shearing is the only way to harvest the wool, as this is the only variety of Angora that doesn't shed naturally.

Rabbit Breeds That Produce Wool

Four Angora rabbit breeds are typically raised for producing fiber.

English Angora. The English Angora weighs 5 to 7½ pounds at maturity and is the smallest of the Angora breeds. It's also the most popular Angora for showing because of its unique face and ear *furnishings* (the long hair on the sides of the cheeks and on the ears). This Angora benefits from daily grooming to keep the coat free of mats and debris, as it has more wool than guard hair by percentage. The fiber from the English Angora wraps very tightly when spun.

French Angora. These guys weigh 7½ to 10½ pounds and are the opposite of the English Angora as far as the wool-to-guard-hair ratio. In this case, the guard hairs comprise more of the coat, making the French a better choice for novice Angora owners. A once-a-week grooming session is all they need. The extra guard hair also allows for more intense colors in the fiber because guard hairs carry most of the rabbit's color. The wool is the undercoat and is normally light-colored or white.

Satin Angora. At 6½ to 9½ pounds, the Satin has a very shiny, crystalline coat. The wool is dense, and its silky texture makes it a pleasure to harvest. Spinners also find this fiber wonderful

Fuzzies and Woolies

Both the American Fuzzy Lop and the Jersey Wooly produce true wool. Since these two breeds are of smaller stature and their adult coats contain many more guard hairs than those of the other Angora breeds, they are less profitable as far as wool production. Still, if you raise Fuzzies or Woolies, their wool can also be spun.

to work with because it's easy to spin.

Giant Angora. This is the largest of the Angora breeds. Weighing in at 10 pounds and up, Giant Angoras are also the best fiber producers in the group. This Angora breed doesn't shed out naturally as do the others, so the wool must always be harvested by shearing.

Special Husbandry Practices

Rabbit breeds that produce wool introduce special concerns.

The most obvious special practice is the almost-daily grooming. This will help avoid matted coats, which can tear delicate skin, and matted hind ends, which can attract flies for egg laying. The rabbits will also ingest less wool, which lessens the chance of a gastrointestinal blockage.

Because these breeds have long coats, it's impossible to tell whether the rabbits are underweight. Angoras need to be picked up and felt to ensure that they are getting the proper nourishment.

Whether wire-mesh or solid flooring, an Angora rabbit's cage needs to be at least spot-cleaned daily to prevent soiled wool and parasite infestation such as flystrike. Flystrike is when flies are drawn to soiled areas of the wool and lay their eggs there. The maggots hatch and quite literally eat the rabbit alive. Flystrike is hard to notice at first because of the rabbit's heavy coat. Once discovered, it could be too late to save the rabbit's life. Because of this, Angora rabbits should be inspected daily.

Again due to fur length, breeding can be difficult. Sometimes a doe (female) needs her tail held up while the buck (male) mounts. Some breeders eliminate this problem by breeding only after harvesting wool.

Angora rabbits ingest a lot of wool while grooming, especially after wool harvesting. When ingested, papaya breaks down the material that holds hair together and helps prevent wool block in their stomach, another potentially fatal malady. You can find papaya enzymes in pill form, or substitute pineapple or pineapple juice, which has the same effect. Many Angora breeders have successfully used papaya to prevent wool block.

General rabbit pellets are not the best food for Angora rabbits. You must use a commercial pellet feed of 18 percent protein that's also high in fiber for growing and maintaining quality wool.

An Angora rabbit's coat can be kept short in cases where a rabbit isn't being shown, isn't pregnant, or doesn't have kits in her nest box (pregnant does or does that have just given birth pull wool from their bodies to line the nest box and keep the babies warm). Keeping the wool short eliminates much of the grooming

How to Groom Angoras

It's paramount that Angora rabbit breeds are kept groomed whether they're for showing, spinning, or just petting. Learning how to groom an Angora can actually mean life or death for the rabbit. Those that become matted are primed for serious health issues such as wool block (see chapter 5) or torn skin. Regular grooming also helps keep dander to a minimum, which means fewer allergens around the rabbitry.

Grooming Tools

- Towel
- Slicker brush
- Wide-toothed comb
- Carpet sample or remnant
- Blower (ideal)
- Grooming table (ideal)

If your Angora rabbit is a pet or is used for fiber production, you'll need a wide-toothed comb and a slicker brush. If you intend to show your rabbit, you'll also need a groomer's high-speed blower. These can be expensive, but if showing Angoras is your hobby, you won't be able to get away with not using one.

1. Put a towel down on your lap for flying fur as well as for the rabbit's security.
2. Place the rabbit on your lap facing away from your body.
3. If you're right-handed, take a handful of wool into your left hand and firmly but gently pull straight up.
4. With your right hand, use the slicker brush to grab at the bottom of the grasped coat; then, taking just a little bit of fur, brush the fur away from the body. Repeat that action, this time grabbing a little more fur than the first time. Repeat until you've gone through the whole handful.
5. Repeat steps 3 and 4 until you've gone through all of the fur. Make sure you rotate the rabbit into a comfortable position for the section you're grooming at the time. For places where there's matting, use the wide-toothed comb first, then go over the section again with the slicker brush.
6. Groom the rabbit's head in the same manner. Pay special attention to the area under the chin, the area between the ears, and the side furnishings. Furnishings can become matted if a rabbit is eating out of crocks (earthenware dishes).
7. Carefully turn the rabbit over by gently but firmly holding onto the ears and the scruff (loose skin) of the neck at the same time.
8. Place the rabbit's ears between your knees, but be careful not to pinch

them. Also watch where your face is while you're leaning over the upside-down rabbit. Rabbits are notorious for kicking and can scratch your face with their toenails.

9. In addition, many breeders find a special rabbit grooming table a worthwhile investment (as opposed to only using their lap) to help them complete their grooming.

Using a Blower

If you're preparing for a show, you'll need to break out a groomer's blower. It's strong enough to put a stop to the beginnings of future mats (called *webs*) in the coat and blow them up the hair shafts until they come to rest on top of the wool. You'll need to use the blower on the rabbit a couple of times a week to keep the coat in optimal condition. Most show Angoras have been introduced to a blower as young rabbits, so don't be afraid use it around them. If they've just been introduced, don't worry—they do get used to it, and in the long run, it is one of the best things you can do for their health.

In a pinch, the cool setting on a regular hair dryer will do. The force isn't as powerful (therefore the results won't be as good), but it's a useful tool anyway. What the blower does is work at the coat like a forceful hairbrush, but it doesn't rip out extra wool the way a slicker brush or comb can. This is important if you're showing your Angora rabbit: you don't want to remove the all-important "density" of the coat.

You use the blower in the same way that you use the slicker brush—by working small sections at a time to get the job done. While blowing, you'll notice a very fine dander wafting out. The white stuff blows everywhere (even all over the groomer), so be aware of this when you choose the location in which to groom the rabbit. And remember: the more the rabbit is groomed, the less dander it will produce.

After the rabbit's coat has been blown out, you can very gently run the slicker brush across the top of the coat using short strokes to remove the webbing that has been pushed out to the tips of the fur.

and other concerns brought on by having a rabbit in full coat.

Rabbit Manure as Garden Gold

One terrific small livestock by-product that is finally getting the attention it deserves is rabbit manure. It's a top-of-the-line, effective, nutritionally rich, balanced organic fertilizer and soil conditioner.

Unlike most animal manures used as soil amendments, which need to be composted—aged and broken down by microbial and macrobial organisms—before being applied to soil, rabbit manure can be used fresh from under the rabbit cages, incorporated directly into the garden and landscaping. Because its nutritional content is already extremely well balanced, it doesn't need the time other fertilizers take to decompose to a workable quality; it can be used right away, without any fear of burning plants. Rabbit manure can be worked into garden soil or even left on top to naturally compost (though manure left on top of soil can attract flies, which are never welcomed by your long-eared pals). Either way, it will improve the texture and tilth of the soil.

Remember that because straight manure hasn't been composted, there's always the possibility of introducing pathogens to humans. So, many gardeners opt to compost their rabbit manure before applying it to their vegetable gardens. In fact, composting rabbit manure with other organic materials will extend its use and help your compost pile break down faster.

Another way to use rabbit manure as a fertilizer for plants is to liquefy it. Fill a 5-gallon bucket with water and dump a coffee can or shovelful of rabbit manure into it. Mix everything up until much of the manure is dissolved, or let it sit for a day. Then water your garden beds with the manure "tea" (again, steer clear of the vegetable bed unless the manure has been composted first).

While rabbits can be very good for vegetables, eating too many vegetables may result in stomach complaints for rabbits.

Rabbits and Red Worms: Perfect Companions

Another way to compost rabbit manure is to use vermicomposting, which is when organic matter is broken down using primarily red worms, or Red Wigglers (*Eisenia fetida*). Red Wigglers are those worms that live within the top 8 inches of the soil. They thrive in damp areas that are rich in organic matter—and they adore rabbit manure. You can keep worm boxes or bins directly under the rabbit cages.

When passed through the worm's gut, organic matter (rabbit droppings) is processed into castings (worm droppings), a nutrient-rich garden amendment. The result is high-quality, vermicompost for your garden. And as a bonus, you have worms to use the next time you go fishing. Rabbit-raisers have been known to sell the vermicompost and worms for some extra profit, too. Most importantly, worm farming in your rabbitry allows you to spend less time cleaning rabbit cages, eliminates manure piles, minimizes odors, and reduces the presence of flies.

Maintaining Worm Boxes

Maintaining worm boxes is simple: it's all about keeping them moist and avoiding excessive urine. Worms can't do anything with dry rabbit manure, so you want to make sure the worm boxes stay damp (not soaked). You'll need to give the box a spray of water every couple of days, but in the summer months, it may be more often.

Worms can tolerate the salt in rabbit urine as long as it isn't highly concentrated. The newspaper strips should soak up some of the urine, but keep an eye out for build-up in the bin. Heavily soaked areas are easily broken up with a small hand rake and mixed in with the rest of the manure. Or you can scoop out the urine-soaked section and toss it into your traditional compost pile; it'll be well received there.

A couple of times a month, use a pitchfork or a small shovel to gently loosen all the contents in the bed so things don't get packed down. Soon, you'll have piles of rich soil in your worm boxes. That's when you know it's time to let your garden beds enjoy the nutrient-rich vermicompost.

Raising Rabbits for Meat

If you're interested in raising livestock for meat production and you don't have a lot of space, there's no better

Vermicomposting is a great way to turn the unwanted into something you'll definitely desire—tasty and beautiful garden greens.

DIY Vermicomposting

Vermicompost is fantastic for the garden. Placing worm bins or boxes under rabbit cages is a popular technique practiced by many rabbit breeders. You can create an entire vermicomposting system using only rabbit manure, worms, and newspaper strips. You'll need:

- A worm bin or box made of plastic or wood, 1 to 3 feet deep (see instructions) and 4 to 5 inches wider than the cage or hutch (so that the box will catch everything that falls out of the cage).
- Half a pound of Red Wigglers *per square foot* of worm bin. Red Wigglers—not night crawlers—are best for composting rabbit manure. These red worms have a voracious appetite for herbivore droppings.
- Worm bedding such as newspaper torn into strips, shredded leaves, straw, shredded documents, or seedless hay.

Before adding any worms, add about 3 to 4 inches of bedding to the bottom of the box, and moisten the bedding with water. When the rabbit droppings are about 1 to 2 inches high, mix the bedding and the manure together, then water it down until it's thoroughly wet but not soaking.

You may need to sprinkle water in the worm boxes every two or three days. You'll want to put your hand into the box every few days to feel if it's getting hot inside. (You could feel the top of the composting material and you'll probably have a good idea of the temperature. But digging in just a little deeper will tell you the full story.) If you do feel heat, mix everything up until the box cools down a bit (worms like it cool). Then feel it the next day—the bin should be cool. Now, place the worms inside the box.

The boxes can be as deep as you'd like, but make your decision based on whether you'd like the boxes to be portable or not. Permanent boxes can be quite deep (up to 3 feet). However, if you'd like to move those boxes when they're full of rabbit manure, you'll want to stay closer to 1 foot deep.

A variation on the theme is to create only a box frame below the hutch—no bottom. This is a great way to take advantage of your local red worms. It's the build-it-and-they-will-come theory. Once the rabbit manure starts hitting the ground, the worms will start migrating toward your box. However, I would still recommend adding a pound or two of worms to get off to a good start.

If you have rabbit cages or hutches with a solid bottom, you can still vermicompost with your rabbit manure. Simply keep a portable worm box nearby along with a handheld scoop. Shovel any rabbit droppings and spilled pellets into the worm bin at regular intervals (the food pellets are just alfalfa, which the worms will enjoy).

candidate than the rabbit. Rabbits don't require true acreage, and most areas don't have heavily restrictive zoning laws regarding their keep. In addition, starting up a meat rabbitry involves very little capital. It's remarkable how many rabbits one buck and two does will generate in half a year, thanks to a short gestation period (a mere thirty-one days). And the amount of meat this small livestock provides is a lot less expensive to produce than that of larger livestock, such as beef or swine.

Rabbits also have little or no impact on the land. Because they're kept in cages and maintained by people, the land is left untouched. Even if they were loose and kept in a colony, rabbits don't have hooves to compact soil like horses and cattle do, and they don't overgraze land the way sheep and cattle can. Add the fact that their manure is the ultimate ingredient for compost, and they're often touted as the "greenest" of livestock animals.

Rabbit meat is all-white meat that's higher in protein, lower in cholesterol, and—with virtually no fat—leaner than more common meats such as chicken, beef, and pork. It's also a valuable source of vitamin B_{12}, selenium, iron, and zinc and is easily digestible. In fact, the United States Department of Agriculture (USDA) has stated that rabbit meat is the most nutritious meat known to man.

Another advantage to adding rabbit meat to your diet is that this livestock is raised off of the ground, so contamination is rare and worms are uncommon. Routine worming (administering anti-worm medication) and the addition of hormones are common practices with traditional livestock. The rabbit is a unique livestock option that rarely needs either treatment.

Common Meat Rabbit Breeds

Like rabbits kept for other purposes, there are certain rabbit breeds that are more conducive to meat production than others. What makes a good meat

Rabbits only rarely require worming or hormone supplementation to raise into healthy, mature adults.

Good Animal Stewards

Whether you're raising rabbits for show, fiber, meat, or companionship, every animal on your hobby farm deserves to be treated with care, kindness, and respect. That is every living creature's birthright.

ADVICE FROM THE FARM

A Match Made in Heaven

Our experts share which characteristics drew them to the breed of their choice.

A Sense of Home

"I spent most of my 4-H time split between New Zealands and Mini Rexes. I loved my Minis; they were my 'babies,' so to say. But something was missing. I found myself drawn to big, lazy breeds. The first time I felt 16 pounds of Flemish Giant in my arms, I was in love, and the versatility and personality of the breed only compounded that.

I love that I've found a breed with the personality to be a pet, that has a wonderful atmosphere for showing, but that also has practical uses for meat and manure. I'll never raise a small breed again. Flemish Giants feel like home."

—Kelly Rivard

Good Hoppers

"As far as the sport of rabbit hopping, all breeds of rabbits seem to do well. They all react differently, and that's what makes it so much fun. Some little, tiny Netherland Dwarfs can hop and jump as high as the Mini Rex and other rabbits twice their size. There are more agile breeds, like the Mini Rex or the Dutch, and then there are the wacko, unpredictable hoppers, like Hollands. They can be all over the place and take in good times with no faults. The kids love them and so do a lot of adults. But there doesn't seem to be one breed that stands out more than any other."

—Linda Hoover

Simply Stunning

"Our family was drawn to the Satin breed because of its dual purpose. It is both a meat breed and a fur breed. We were drawn to its amazing fur and beautiful colors, with the Red variety drawing us in first. The Red Satins have stunning, shiny color combined with that luxurious Satin texture that makes this a favorite breed among many. It is the 'eye candy' among many breeders."

—Jennifer Ambrosino

Love at First Lick

"I didn't choose my breed—my breed chose me. I got into Mini Rexes because my 4-H leader had her very first litter of Mini Rexes at her barn. When I went to pick up the broken red Mini Rex from the cage, I brought her up to my face, and she licked my cheek. From then on I knew I was going to buy her. Now I have twenty-three Mini Rex rabbits, and they all started from that cute, broken red Mini Rex that I named Butter."

—Kendyl Schultze

Just a Little Thing

"One of the most appealing aspects of raising Netherland Dwarfs was the fact that they require little food. A full-size Netherland Dwarf eats no more than 2 ounces of pellets a day. Some breeds can require triple that amount of feed, which means they cost more money to raise. The Netherland Dwarf is also one of the most popular and commonly seen breeds. While many breeds only come in a few colors, Dwarfs have more, including orange, chinchilla, broken, tortoise shell, otter, and many others. They're also very distinct due to their massive, round heads; short, thick ears; and big, bright eyes. Netherlands are generally sweet-tempered."

—Brenda and Sarah Hass

Color Me Happy

"Fifteen years ago, I fell in love with the fur of the Mini Rex. They were the perfect size for me as a beginning 4-Her. I love the challenge of a breed that combines type, a unique coat, and color. The variety of recognized colors of the Mini Rex allows everyone to find one that they enjoy. This breed has a relatively short 'shelf life' as far as showing, and they are very competitive, which creates a challenge."

—Keelyn Hanlon, ARBA Judge

Compare Calories per Pound

Meat	Calories
Rabbit	795
Chicken	810
Veal	840
Turkey	1190
Lamb	1420
Beef	1440
Pork	2050

rabbit? Considerations are really about economics. *Feed conversion ratio* (how much feed is required to produce one pound of meat) is important, as is reproductive efficiency.

It may seem to make sense that a very large breed of rabbit produces a lot of meat. However, many breeds aren't reliable procreators and are difficult to get pregnant, which can throw the economical balance off. The *dress-out percentage*—or how much edible meat is actually produced (the hide, bones, entrails, and so on) from the rabbit—is another factor that's considered when looking for the perfect meat rabbit for your hobby farm. Of course, there can be other individual considerations, such as your climate or if you're also showing or selling the rabbits for fancy.

Here's a list of the most common rabbit breeds raised for meat; it's by no means exhaustive:

- Beveren
- Californian
- Champagne D'Argent
- Florida White
- New Zealand
- Palomino
- Rex
- Satin

Housing and Feeding Rabbits

Housing and feeding rabbits isn't complex in the least. It does, however, require some common sense on the part of the person running the rabbitry. In this chapter, we'll discuss the many ways to house your rabbits, and I'll help you decide which is best for you. And, of course, your rabbitry should always be supplied with quality high-fiber food and fresh water. But does your rabbitry need toys? Do your rabbits need those little salt spools? Let's go over the specific considerations of housing and feeding so that you can make the best decisions for your furry charges.

The Rabbit Residence

Where your rabbits will call home is one of the most important decisions that you'll need to make. The following considerations may sound like a lot at first, but as I said earlier, all you need is a little common sense to get it right.

Placement and Protection

If you're starting from scratch, the most logical place to begin is to decide where the rabbitry should be situated. One of the most important housing issues you'll face is safeguarding your rabbits from the weather. Because rabbits won't survive direct sunlight and aren't especially tolerant of heat, it stands to reason that they should be kept on the northernmost side of the house. You'll still need to provide protection from the sun, but this will be easier to do in a cooler area.

Even if you live in a cooler climate, you should err on the side of caution, as most cool places still gather some brilliant sun in the summer months. Rabbits do best in temperatures between 50 and 75 degrees Fahrenheit. And as long as they're protected from rain, snow, and driving wind, they'll be fine in lower temperatures. But even in cooler temperatures, rabbits can't survive the summer sun bearing down on them. In fact, heat is the toughest thing for them to deal with. At more than 85 degrees Fahrenheit, rabbits begin to get uncomfortable even in the shade. When the temperature rises, place a freestanding water mister or fan near the hutch or a frozen water bottle in the cage. If the rabbit continues to pant, and you don't have fans or a mister system, the rabbit should be brought indoors. If you live in an area that sees frequent high temperatures, and you don't want bunnies running around your home for the summer, be prepared to have one of these systems in place.

When deciding where to place your rabbitry, also think about what you'll need to make it run. If you don't want to haul a hose or an extension cord a long distance, place your rabbitry near utilities. But if you're going to have electricity running to your rabbitry, make sure that you don't store your hay anywhere near the outlets or lightbulbs. If you're really concerned about potential fires, you can always add a smoke detector to your facility's decor.

When I talk about a "rabbitry," I'm referring to the large structure that houses the smaller cages. Rabbit owners use a variety of buildings to house rabbits and their cages. The structure need not be elaborate; a simple lean-to can suffice. But if you use a structure that isn't fully

Acquiring Cages

Some rabbit keepers make their own all-wire cages or hutches because, in the long run, it's a major wallet-saver. But for those who don't mind the price and want cages now, they can, of course, be purchased. Some places are cheaper than others, so shop around for the best deal.

- **Local pet store.** This is, hands-down, the most expensive place to buy cages. Still, in a pinch, you'll find them there.
- **Local rabbit breeder.** Many rabbit breeders not only sell rabbits but also have a side business of selling rabbit supplies. The prices here vary. The cost will mostly depend on whether the cages are manufactured or whether they're handmade by the breeder.
- **Online or mail-order rabbit-supply companies.** This is usually the least expensive way to purchase rabbit cages, especially if you're looking for more than one or two. The cages come to your door partially assembled and aren't difficult to assemble fully.
- **Rabbit shows.** This is one of my favorite places to purchase rabbit cages; they're usually assembled and priced decently. Many of the companies at these shows are the same ones that you would order cages from online.

Although these guidelines are usually true, don't overlook the occasional sale from any source. Researching all of these sources before buying cages can really pay off.

This hutch setup is convenient for its owner because it's near the house, but it's also convenient for the rabbits because they have both protection and a little bit of freedom.

contained, then you'll need something that can be pulled down over the front of the cages to protect them against sun, rain, wind, and snow. Some rabbitries have very angled rooftops that can help, but usually something else, such as a tarp, must be attached to the structure for added protection.

In addition to weatherproofing your rabbitry, it's vital that you provide protection from predators. The ones that initially come to mind are raccoons, foxes, weasels, and other wildlife. But make no mistake, the "friendly predator" that most often causes damage and fatalities in the rabbitry is the family dog—not just yours, but neighborhood dogs, as well. The key is to secure your rabbitry against any and all possible predators to prevent heartbreaking events.

You can provide protection one of several ways. For one, you can house

What's a Four-Hole?

You may hear someone say they have a four-hole for sale. This means that they have a cage that can house four rabbits, although it doesn't say anything about the dimensions of each hole or cage. A phrase often overheard at rabbit shows is, "I just bought another doe—I had an open hole." This is because rabbits are like potato chips; rabbit-raisers tend to get itchy if they have an empty cage at home!

your cages inside a small barn or shed that has doors. Another option is to install a chain-link dog run that has a gate and top around the rabbitry. A garage or mudroom can easily be turned into a rabbitry. Or, if the cages are inside a simple plywood outbuilding, you can attach a cover to the front of the roof so that it can be rolled down and secured for the night. Removable sides for the cages themselves are also a good idea. I know a couple of rabbit keepers that even have removable wooden panels for their hutches.

The wire cage is a standard in the rabbit-keeping hobby.

Cage Choices

When it comes to rabbit cages, you have more choices than you might think. First, you have the traditional rabbit hutch built with wood and welded wire. (By the way, don't ever use a hutch that's been built with chicken wire; it's much too hard on rabbit feet and it's easily broken by predators and rabbit teeth. See the sidebar "Rabbit Feet and Welded Wire" on page 48.) People generally like the look of rabbit hutches, and you can paint them to match your home or with whatever colors you like. You can also purchase removable wooden fronts to keep out inclement weather. The down side is that the rabbit inhabitant will probably chew on the hutch and soak the wood in urine if it isn't thoughtfully constructed so that the rabbit can't easily reach the wooden frame. Hutches can also be costly to build or purchase depending on how many you need. On the other hand, you can easily find used wooden hutches online or in your local paper.

You can opt for the all-wire "hutch" or cage that omits the wood. Many of these cages come attached to another cage. Some of them have slide-out litter trays, and some don't. One of the most popular wire cage arrangements today is called a "stackable." These all-wire cages are stacked three to

How Much Cage Space?

The rule of thumb for cage space is 1 square foot of space per pound of rabbit. So, a 6 pound rabbit needs a cage that's at least 2 feet wide by 3 feet long (6 square feet) to be comfortable. I personally prefer to offer a bit more room than that. Most cages are 18 inches high or a little taller, but for the smallest breeds, such as Netherland Dwarfs, you can purchase shorter ones.

four cages high on top of one another and secured by metal legs. Each stacked cage has an individual pull-out tray. Stackables take up the least amount of space in the rabbitry. There are also rabbit cages with solid bottoms, which are especially nice for rabbits living inside the home, but they work just as well outdoors, too.

Rabbit cages can also be constructed on the ground as long as there's some sort of barrier, both to keep them from digging out of the cage as well as to keep anything from digging into the cage. Wire cages sitting on bare earth with something like straw or bedding on the floor of the cage (on top of the wire) works well. One of the drawbacks to having rabbits on the soil is that they are more likely to contract worms. Most rabbit breeders only occasionally have to deworm their herd because they use elevated cages. Rabbits that are kept in contact with the ground, however, may need to be on a deworming program.

Build Your Own Cages

If you're planning on keeping three or four rabbits, cages are probably the least expensive way to house them. However, if you're planning on keeping a large rabbitry, or one that you know will grow, your wisest choice may be to build the cages yourself. Purchasing the building materials in bulk can be the least expensive way to go.

Another great reason to build your own cages is that you can design them yourself, taking advantage of every square foot of space in your rabbitry. Once you're good at it, you can also create a side business for yourself selling cages to other rabbit enthusiasts. It's something to think about: selling just a cage or two a month can go a long way in offsetting your rabbitry expenses. Building your own rabbitry also creates a wonderful sense of accomplishment.

Sheds and Barns

You can use a potting shed, barn, lath house, garage, or any other existing outbuilding to house your rabbit cages. If you don't have something readily available, you can always build a shed structure specifically for the cages—

Not every cage works for every rabbit. Create a setup that works for you and your charges.

Rabbit Feet and Welded Wire

Welded-wire cages have been the cause of a painful condition called "sore hocks" in some rabbits. This has certainly been an issue for some rabbits with very little fur on the bottoms of their feet. The general feeling is that rabbits are more comfortable with more support underfoot.

Removable, hard plastic mats are a terrific solution to this problem. These mats are easily secured to the cage floor and have thin slits that allow manure and urine to escape through them. With this flooring, the rabbits are happier and their owners can continue to use the well-ventilated and easy-to-clean wire cages.

That said, wire flooring isn't the only way that rabbits acquire sore hocks. Rabbits sitting in their own urine and feces will also bring on this affliction. Solid flooring will only prevent sore hocks if it is cleaned thoroughly and regularly.

something just large enough for the cages to slide into is sufficient. You'll want a shelter that's weatherproof and critter proof but with plenty of ventilation so that your rabbits don't suffocate or overheat. Rabbitries can be as simple or as elaborate as you'd like them to be. Some of the more lavish rabbitries boast concrete floors (easy to hose), fans, heating systems, air-conditioning, insulation, sinks with running hot and cold water, supply cabinets, built-in grooming tables, and fly-spray mister systems.

The Importance of Ventilation

Ventilation is often overlooked when creating a rabbitry, yet it's one of the most important aspects affecting your rabbits' health. All wire hutches have ventilation in spades, but whatever structure houses the cages needs to have it as well. Many prefabricated sheds or barns have to be modified to add additional ventilation. Placing strategic air vents in the rabbitry will allow the breezes outside to blow through, creating better air circulation. If your building needs a little help despite good air ducts, add some fans. If you use fans, you'll need to have electricity in your rabbitry—an excellent idea in any case because lights are always helpful in the winter. But if your rabbits are right next to the house, you may be able to get away with an extension cord from an outside electrical outlet.

Extreme Weather

Rabbits tolerate cold temperatures much better than they do hot temperatures. In fact, while it is nice to have some heating in the rabbitry, even in the coldest areas of the country, you don't need to heat it much. As long as the rabbits are protected from wind, rain, and snow, they'll do very well.

Heat is another story. Each rabbit will tolerate different temperatures, but generally around 85 degrees Fahrenheit, rabbits get stressed and become visibly uncomfortable. To ward off potential heat stroke, some breeders freeze 2-liter bottles filled with water and place them in the cages for the rabbits to rest against. However, some rabbits never get the hang of leaning against the bottles, so regular fans, misting fans, garden-

As many children might, this little guy chose to spend his playtime in the snow.

Wire Primer

All wire is not created equal when it comes to rabbit housing. Whether you're purchasing or building cages, there's a few things you should know about wire.

Wire specifics. The wire used for rabbit cages should be galvanized welded wire. If given a choice, choose wire that was welded *before* it was galvanized; your rabbit will be sitting on smoother and stronger wire. Note: hutches made with chicken wire or hardware cloth are *not* suitable for housing rabbits!

Wire for the sides and top of the cage. With wire gauges, the higher the number, the lighter the wire. That means that 14-gauge wire is heavier (thicker) than 16-gauge wire. You can use 16-gauge wire for small rabbit breeds. Medium or large rabbits need 14-gauge wire cages. (These will last longer, too.) The mesh (the rectangular holes) should be 1 inch by 2 inches, or possibly a little smaller.

Wire for the cage floor. Again, you want 14-gauge galvanized welded wire. But, for better support, the mesh should measure ½ inch by 1 inch. This mesh size will allow manure to drop down to the tray or ground below. Always use smaller mesh for the floor of the cage than you did for the sides. Your rabbit's feet can break or become dislocated if they get caught in larger mesh.

Baby-saver wire. This is a special wire that's used for the sides of cages housing a breeding doe. It's expensive, so expect to pay a little more for cages that are built with it. The wire is very efficient at keeping kits inside the cage should they accidentally come out of the nest box while they're still young. The bottom of baby-saver wire has ½-inch by 1-inch mesh while the top graduates to 1-inch by 2-inch. If you're breeding rabbits, it's worth the extra expense.

Keeping Rabbits Indoors

If you plan on raising only a couple of rabbits, keeping them inside the home with the family has some nice advantages. One is that indoor rabbits have the opportunity to play more often with their families. This makes it much easier to raise a friendly rabbit and enjoy its company. Another plus is that, aside from bringing your rabbit outdoors to play, you don't have to worry about protecting it from predators (although you'll still want to keep an eye out for the family dog and for small children who don't understand how to handle rabbits). You also don't have to worry about extreme weather if you keep your rabbits inside. If you're comfortable, they should be, too. Indoor rabbits may live longer than their outdoor counterparts thanks to this lack of predator and weather worries. Not to worry, though; if your exterior rabbitry is well planned and protected, outdoor rabbits live every bit as long as indoor ones do.

Indoor rabbits are sometimes housed in a cage at night and allowed to roam free in the home during the day. Some are only brought out of their cages to play. And some—because rabbits can be litter box trained—live cage-free inside the home. However you decide to keep your indoor rabbit, you need to take some precautions before you allow your rabbits to run around the house.

Rabbit-Proofing Checklist

- Relocate plants to a nonrabbit room; many are toxic to rabbits. If not, eating too much of any plant can still give rabbits diarrhea and an upset stomach.
- Get down on all fours and look for any hidden nooks and crannies that your rabbit might be able to squeeze into and become stuck or lost. Domestic rabbits are burrowers by nature.
- Find all the wires and cords the rabbit could chew, and either relocate them or cover them with aquarium tubing. A rabbit can electrocute itself or severely burn its mouth by chewing on wiring. Rabbits have to chew in order to keep their ever-growing teeth filed down.
- Lock up all toxic chemicals so they can't be reached. This includes antifreeze, cleaning supplies, pesticides, and fertilizers.
- Place things such as trash bags and clothing out of your rabbit's reach. Should a rabbit ingest fabric or plastic, it can easily develop a gastrointestinal blockage.
- Choose an uncarpeted room for it to play in. The next best thing to chewing that a rabbit can think of is digging. It is part of their genetic makeup. Carpet fibers are notorious for causing intestinal impaction. Rabbits may also scratch at doorways, so many indoor rabbit keepers attach Plexiglas to the lower part of doors to prevent door damage.

A frozen bottle of water in this rabbit's cage ensures that it stays cool in the summer heat.

hose misters, and overhead tube misting systems are extremely useful. (See the information on page 85 of chapter 5.)

Other Residential Necessities

Now that you've finished the main structure, you'll need to consider the incidentals. Some of the suggestions below are optional, but consider that they each offer something for the rabbits' health or happiness.

Food dishes. Ceramic crocks can work well for holding pellet feed, but rabbits may flip them over, so the best crocks are the heavy ones. Crocks also take up precious cage space, and rabbits may sit in them and end up with manure and urine in their food. For pellet feed, I prefer a metal self-feeder called a J-feeder that attaches to the cage from the outside with only the bottom lip sitting inside for the rabbit to feed from. The bottom of the feeder has wire mesh so that the dust from the pellets falls through. This is nice because the pellet dust tends to get caked in crock feeders. In addition, the

J-feeders hold so much feed that you'll find yourself refilling them less than you would using crocks.

Water containers. You can offer water in a crock or in a water bottle, which attaches on the outside of the cage with only the nipple of the bottle sitting inside. If you have a large herd, you may want to use an automatic watering system. A watering system has plastic tubing that runs through the rabbitry and connects to every cage, where small nipples sit inside the cages (similar to the water bottles). I like using individual water bottles over a watering system because I can tell at a glance if my rabbits are drinking enough water or not. A rabbit that isn't drinking may be ill, or the nipple may not be working properly.

Hay rack. A hay rack is nice to have inside the cages; it keeps hay fresh and off the floor. This is an optional piece of equipment.

Hiding place. Although it's not necessary, rabbits feel more secure when they have a place to tuck themselves away that's reminiscent of their wild ancestors' warrens. These havens can be made of wood, cardboard, or other rabbit-safe materials.

Urine guards. Many cages that are sold preassembled have what are called "urine guards." These are 4-inch metal strips that run along the inside of the cage to keep urine inside the cage and not on the floor or in the surrounding area. It's important to note that while urine guards can help keep babies from falling out of the cage, they aren't as reliable as baby-saver wire (see the sidebar "Wire Primer" on page 49).

Toys. Some rabbits enjoy having chews and toys in their cages, and most of these don't have to be purchased. Toys that double as gnawing items are nice because rabbits' teeth are always growing and these types of toys will help keep them worn down. Try those dangly wooden toys that you find in pet stores. Nearly anything safe to chew and ingest is fair game: toilet-paper rolls stuffed with hay, apple tree branches, cardboard

Cloth Caution

Don't put towels, blankets, pillows, or anything cloth inside your rabbit's cage. Rabbits will chew on those, too, but they can't vomit them up. This can cause a gastrointestinal obstruction in the rabbit that can only be rectified by surgery. Unfortunately, most rabbits don't make it to surgery before it's too late.

Rabbits need some downtime and exercise, too, so make sure you give them the opportunity.

boxes, and grass mats and huts. When you notice the paper products becoming soiled, you should toss them.

Play yard. Play yards for rabbits are the same as the ones used for dogs. They're portable fencing structures that allow the rabbit a little freedom and a change of scenery. I love having play yards around because it allows me to give my rabbits exercise while still keeping them confined in a safe area. I don't have to worry about them eating the plants in my yard that may be dangerous or about predators having easy access to them. I also don't have to worry about having to chase them down, which only scares them and pushes us backward as far as trust in handling. When you use a play yard, always place part of the yard in the shade or position an umbrella over half of it. Be aware of the sun shifting as the day goes on. Don't leave your rabbits unattended.

Running the Rabbitry

The first step to enjoyable and successful rabbit keeping is to have a sound management program. As the manager of your rabbitry, you decide what program will work best for your situation. Here are some general tasks that should be incorporated into running the rabbitry:

The presence of dangerous and clever predators is a really good reason to make sure that your rabbits are secure and supervised outdoors.

Little Helpers

If you have family members who can help with rabbit care, by all means, incorporate them into the program. However, be aware that if the family members are children (even teens), they'll ultimately need to be guided. This usually means that, for the sake of the rabbits, you'll need to double-check that everything is running smoothly.

Daily Management

- Replenish food and water every day. If you're using water bottles or crocks, you may have to refill the water twice a day in hot weather.
- Look at every rabbit, keeping an eye out for problems or health issues, such as unusual behavior, loose stools, watery eyes, or runny noses.
- Handle the rabbits daily if possible. Your rabbits will be friendlier and you can keep a closer eye on their body condition.
- Check on any litters if you have does raising kits. Lookout for illness and change wet bedding in the nest box.
- Spot-clean cages with a handheld scoop. This simple task makes weekly cage cleaning simple and fast.

Weekly Management

- Handle your rabbits. If you aren't handling them daily, at least do it weekly. Check their teeth for malocclusions or breaks.
- Make repairs to cages and other equipment. You'll thank yourself later for staying on top of the small things.
- Clean all cages weekly (at least). This is one of the best ways to raise healthy rabbits—and it keeps the fly population down.
- Check your calendar or hutch cards (see the sidebar "The Hutch Card," opposite) for does that are preparing to kindle (give birth) in the next week so that you're prepared.
- Thoroughly clean nest boxes that are removed from cages, or add nest boxes to cages with does that'll be kindling in the next week.
- Check your supplies so that you're aware of when you will need more feed, hay, bedding, and so on.

Monthly Management

- Give the entire rabbitry a once-over. Do you need to add fans or misters for expected hot weather? Extra straw for bedding for a cold front?
- Check the litters again. Which litters have outgrown the nest box? Is it time to separate them by sex? Any runny eyes or noses?
- Check your first-aid kit and restock whatever you've used recently.
- Trim every rabbit's toenails and check every rabbit's teeth.
- Fill out pedigrees for new kits, update hutch cards, and update your expense book.
- Make sure the calendar is marked for upcoming shows and that you have everything you need to prepare for those shows.

The Importance of Record Keeping

For rabbit keepers who have only one or two rabbits, record keeping may seem like overkill. After all, it's fairly easy to remember the details on just a couple of rabbits. But as your rabbitry grows, you'll be amazed at just how easy it is to confuse sexes, ages, and breeding dates. Specific details are also extremely important if you plan on making money with your rabbits. Keep accurate records of:

- Breeding dates and which doe was bred to which buck.

The Hutch Card

In rabbit keeping, the hutch card is attached to the front of a rabbit's cage, and it contains current information pertaining to that rabbit, such as the rabbit's name, age, breed, which buck bred the doe, litter due dates, how many kits were born, how many kits survived, and so on.

You may purchase preprinted hutch cards from rabbit product catalogs, obtain them free from certain feed companies, or make them yourself by hand. Although I think hutch cards are very handy for on-the-spot info, I think of them as secondary and prefer to keep records in a hard file or on a computer program. Here's an example of a hutch card:

Name:
Sex:
Birth Date:
Breed:
Ear #:
Bred to:
Date Bred:
Due Date:
Number of Kits Born:
Number of Kits Weaned:
Notes:

- Birthdates of litters born (transfer those dates to individual rabbit cards *on the day* they are separated).
- Pedigrees for each rabbit (fill them out completely).
- Every expense in the rabbitry and any income from your rabbits.

Keeping accurate records allows you to have insight on which rabbits are producing the highest quality litters, which rabbits do the best at shows, which pairings work and which don't, and how much you're actually making with your endeavor. The current direction in rabbitry record keeping is the use of computer programs made especially for hobby farmers. These programs allow you to keep track of your rabbits' lineage, tattoos, show records, pedigrees, and your own business expenses. Many rabbit keepers feel that this is the best way to go as far as accuracy and simplicity when dealing with rabbits on a large scale.

This rabbit's clean cage and fresh shavings ensure its health and happiness.

Controlling Flies in the Rabbitry

Flies are attracted to rabbit urine and manure, and they show up practically overnight once a rabbitry is set up. It's important to control the flies in your rabbitry, and it's something that's best addressed before you bring home your rabbits so you can launch a preemptive strike. Flies aren't just a nuisance for you and your neighbors—they also can spread disease. A rabbit with loose stool is a target for fly strike, which is almost a sure thing if you have a fly problem. (Fly strike is when a fly lays eggs on a rabbit. When the fly larvae hatch, they eat the rabbit's flesh. Not a pretty picture! See chapter 5 for more information.)

The first way to keep flies to a minimum is to clean the cages as often as possible. It's perfectly fine to simply spot-clean in between full cleanings; you just don't want the manure piling up. The best place for rabbit manure is the compost pile, but try to mix it in rather than just drop it on top of the pile. This will help keep flies from hovering around the compost pile. As we discussed in chapter 2, red worms can also help decompose the manure for you, creating wonderful compost for the garden in the process.

For cages that have no drop pan underneath, if you're not raising red worms directly under the cages, you can spread horticultural lime there; this is said to make the ground inhospitable as a breeding area for flies. Remember that you're looking for the *horticultural* type of lime, or "barn lime," which is gray in color—not *hydrated* lime, which is pure white and not safe to use in this manner.

While there are many flytraps on the market, using traps alone won't make a big dent in the fly population. (Not to mention that the traps smell terrible, so they have to be placed strategically or you'll be chased away from the rabbitry, too.) Fly parasites and natural predators are another option; you can order them online. These parasites hatch and kill the fly larvae before they can grow up to become flies. I've found this approach to be helpful, but once again, not as a sole preventative.

You always have the option of using old-fashioned sticky tapes. They work pretty darn well, but dust and debris flying around the air shortens their shelf life—and woe unto the person that gets his or her hair caught in it! Another fly-control system that many rabbit-raisers rely on is a battery-operated dispenser that automatically releases a metered dose of pyrethrin, a pesticide derived from the pyrethrum flower, every so often. This capitalizes on an effective organic gardening method that won't harm your rabbits.

Be a Good Rabbit Ambassador

If your eyesore of a rabbitry or the wafting of strong urine smells annoys the neighbors, they may complain to the authorities. Of course, you want to be on top of your city's latest zoning laws pertaining to rabbit keeping. Even if you abide by all of the city's restrictions, your neighbors' complaints will still cause an upheaval that none of you need. Aside from the fact that you'd like to enjoy a harmonious relationship with your neighbors, you also want to be a model ambassador for rabbit keeping everywhere.

If you live in an area where your neighbors can see the structure, it's in your best interest to make your rabbits' home as attractive as possible. I've seen and created some very charming rabbitries, and I encourage you to give some thought to the finishing touches on your own design. Rabbitries can be painted to blend in with their surroundings or painted to match the color of your home, which creates a look of uniformity. You can also design your rabbitry to look like a dollhouse or gingerbread house or use flower gardens, shrubbery, and trees to create an attractive environment. If you're using lath (latticework built around the cages), plant some climbing roses or other vines next to the lattice for a country garden look.

The best way to achieve peace, acceptance, and even support for your rabbitry is by being considerate of your neighbors. You can do this by not overwhelming your property with livestock, keeping healthy and happy animals, and following excellent husbandry practices.

Rabbit Cuisine

There are as many opinions on feeding rabbits as there are rabbit keepers. While hobby farmers are looking to meet their rabbits' nutritional needs, they're also looking to do it simply and inexpensively. Your best bet is to come up with a plan that works for you and your rabbitry and modify it later if necessary. Keep in mind that you want to provide optimum-quality feed. You don't have to stop at pellets, however. Your rabbit can enjoy grass hays and other goodies, as well.

Water

Water is a top priority when it comes to proper rabbit care. Cool, fresh water is a

You may find that your rabbit is a bit picky about its food. When swapping out brands, make sure to do it gradually and provide time for the adjustment.

must *at all times*. In fact, rabbits won't touch their food if they're out of water. If a rabbit goes off its feed, the first thing to check is the water supply. Is the crock or water bottle empty? If you're using water bottles or a watering system that delivers water through a nipple, check that the nipple is letting the water flow freely.

Commercial Pellets

High-quality pellet rabbit feed should make up the majority of your rabbits' rations. These cylindrical pellets include everything your rabbits need—protein, phosphorus, calcium, and trace minerals from varying ratios of alfalfa and other hays such as timothy or oat. I've found that my rabbits prefer some brands over others due to the quality and freshness of the hay. In a pinch, I've purchased an off-brand feed only to have my rabbits turn their twitchy noses up at it.

A nice pellet feed will look rich and smell fresh. Those of lesser quality will look incredibly light and dry and not have much scent when you stick your nose into the bag. That said, if you choose a pellet feed that's made of timothy, it may be lighter than alfalfa pellets—this is perfectly normal and fine. In fact, timothy pellets will be higher in fiber, which is a good thing. I prefer timothy for that reason. However, you might be limited to what is available in your area (unless you order your food online).

When switching brands or quality of pellet feed, be sure to make the switch gradually. Rabbits' systems are delicate, and even a basic switch can throw off their intestinal balance.

You'll find many formulas on how much pellet feed you should give your rabbits. I follow a general rule of thumb for feeding adult rabbits: give them as much as they'll eat in a day. If you find there are always pellets left in their bowls, they're getting too much. But if they jump on top of your hand every time you fill their bowl, then

Storing Pellet Feed

It's important to keep pellet feed as fresh as possible for your rabbits, and this includes keeping out unwanted critters such as mice and rats. Feed should be kept in a cool, dry place. You can use one of any number of containers on the market. Believe me, I've tried them all, and for me, nothing is more effective (or lasts longer) than the old-fashioned metal garbage cans. Absolutely nothing can chew through them, and they seem to keep feed the driest, too.

you may be offering too little. I've had great luck with this formula, and I never have to measure my rabbits' feed—not even for my show rabbits. Baby rabbits have different needs, which we'll go over in chapter 7.

Certain circumstances will call for a change in portion, a special mix, or an extra nutritional supplement. For instance, you might have to adjust your rabbits' feed when the weather turns cold (if you're not in a temperate area), as they'll need to consume more calories to keep warm. A nursing doe will also need more calories to help produce milk for her kits. A mature rabbit needs less food than a juvenile rabbit that's still growing (regardless of its size). Angoras need a pellet feed that's extra high in protein to produce their long wool. In addition, Angora owners sometimes feed their rabbits special mixes that help pass ingested wool through their systems to avoid a blockage (Angora rabbits ingest a *lot* of wool). You might also give show rabbits a special mix to help them gain a little weight as well as to encourage a nice coat. These are just a few examples of special circumstances you may encounter.

A hay rack can be handy for keeping your rabbit stocked and its cage clean.

Hay

Most rabbit-raisers supplement their rabbits' diets with free-fed grass hay such as timothy. Rabbits are foragers by nature and are healthier and happier when grass hay is added to their daily diet. Most rabbit keepers recommend a filled hay rack for each cage. I always keep oat hay around for my rabbits. Oat hay is like magic when it comes to firming up loose stools, and it makes great bedding, too.

Vegetables, Fruits, and Treats, Oh My!

This is a controversial topic that needs clarification. Some people feel that their rabbit's "natural" diet would consist entirely of vegetables. This simply isn't true. Most wild rabbits aren't sitting in a voluptuous garden, happily munching fresh veggies all season (with the exception of Peter Rabbit, and even he ended up sick and in bed). The majority of wild rabbits forage on grasses and other greens, shrubs, bark, and twigs. They may or may not occasionally run into something a bit juicier in the way of vegetables and fruit. Therefore, a vegetable diet for rabbits is not replicating the diet of their wild cousins.

That said, rabbits are herbivores and can benefit from nutritional vegetables, even on a daily basis. In fact, rabbits can eventually enjoy 1 to 2 cups of veggies per day, as long as the vegetables are introduced very slowly, and they aren't plied with watery lettuces that can bring on diarrhea.

Now we're getting to the biggest problem associated with feeding rabbits too many vegetables too quickly: diarrhea. Diarrhea isn't a disease, it's a symptom, but it can kill a rabbit very quickly. Some rabbits can be cured of it overnight if offered oat hay or *old-fashioned* oats (*not* oats with the added sugar). But some cases are hard to turn around, and the rabbit ends up dying before the diarrhea clears up. Something to note: Don't offer baby rabbits greens or vegetables. They'll have an extremely hard time surviving a bad case of diarrhea. (See page 84 in chapter 5 for more information on diarrhea and its treatment.)

Being an avid gardener, I do enjoy sharing some of my garden bounty with my rabbits, especially because

What about Weight?

Make it a weekly habit to run your hand over each one of your rabbits' bodies. They should feel filled out without having a large stomach or fat pads on their shoulders. You should be able to feel their hip bones, but they shouldn't be protruding or sharp. If you can easily feel your rabbits' ribs and their spines are obvious, you need to beef up the groceries.

If you're raising show rabbits, you'll soon realize that there is a weight limit for each breed standard. Sometimes people will give their rabbits less food to bring their weight at or just below their breed's weight limit. If the rabbits are obese to begin with, this makes sense. But many times, their weight has to do with improper breeding practices. In other words, the rabbit breeder bred two large animals together without thinking about the resulting offspring's adult weight. The breeder then sells the offspring to an exhibitor, and in an effort to not be disqualified from the show table, the new owner ends up starving a rabbit that was simply bred to be larger. If you're interested in showing your rabbits, make sure you purchase animals from breeders with good reputations for show stock, and don't adjust your show rabbit's feed unless its weight is unhealthy for its structure.

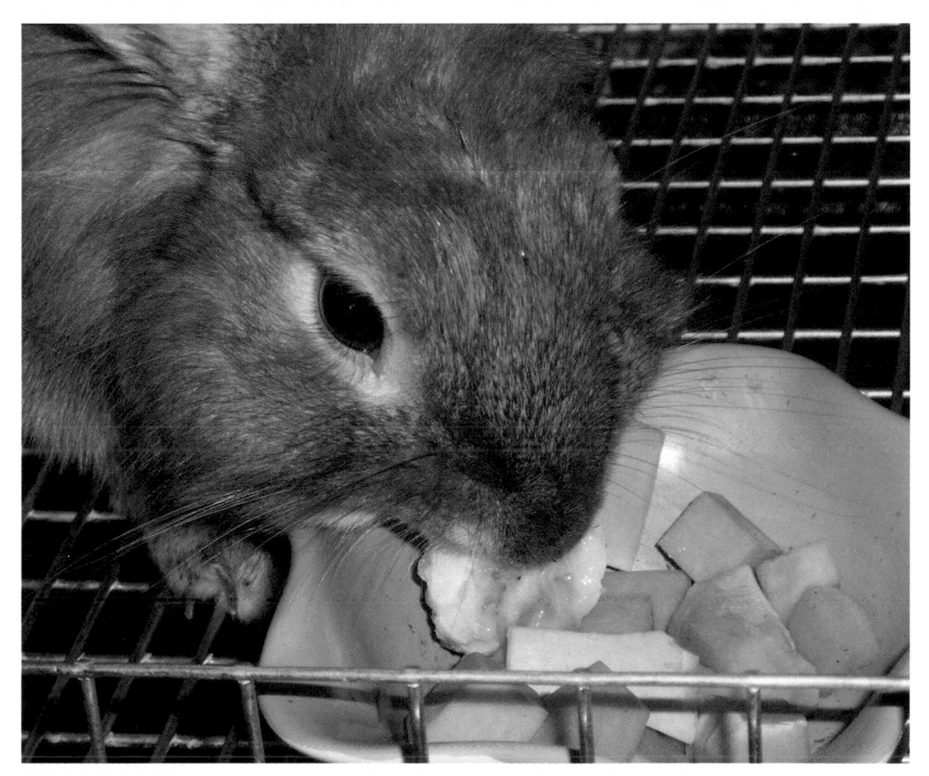

Offer rabbits fruits and veggies—such as bananas and squash—sparingly to save their delicate systems.

their manure helped create it! But I only offer vegetables to my rabbits in very small amounts. I also vary the types of vegetables I offer to maintain a good nutritional balance. Because I'm cautious, I monitor those rabbits with sensitive digestive systems and personalize their snacks. I have one little blue doe named Blizzard that gets diarrhea if she has anything more than a tiny carrot. In fact, she tends to get it so easily that I always add some old-fashioned oats to her food, which works beautifully. If you want to offer a handful of veggies to your rabbits as part of their daily diet, do so *very* slowly and watch for signs of watery stool. If things are getting loose down there, back off a bit with the veggies and greens, and offer the affected rabbits oat hay or old-fashioned oats.

You can give rabbits fruit sparingly. I give my rabbits apple pieces and the like once in a while. The same rules of caution and moderation apply here. However, steer clear of fruits like peaches; super sugary and watery fruits loosen stools almost immediately, so apples and pears are your best bet as far as fruit snacks. Other treats that rabbits really dig are dandelions (if they're from your lawn, be sure that herbicides or pesticides haven't been sprayed on them), dry bread, parsley, carrottops, grape leaves, and plantains.

Why Your Rabbit Is Eating Manure

On occasion, you may catch your rabbit eating manure. People tend to be concerned by this. After all, it does seem like a problem. This behavior is called *caecotrophy*, and it is an important part of the rabbit's digestive process.

Rabbits' bodies produce two kinds of feces. The first one is the hard, round, black (or dark brown) dropping that you typically find in or under its hutch. The second type of feces (called *caecotrope*) is softer, greenish, mucousy, and looks like a cluster of grapes. The rabbit's cecum (a sac that contains healthy bacteria and other organisms to help the rabbit digest food) produces these softer feces, and the rabbit ingests them directly from its rectum. Sometimes you'll see the caecotrope in the cage if the rabbit was disturbed before it had a chance to ingest it. Rabbits use the microorganisms in the caecotrope to help them digest their food properly. This is a nutritional requirement for rabbits; without it, they become malnourished.

Rabbits will munch on pretty much anything they can get their teeth around—especially if it's green. Supervision and control are very important when it comes to a rabbit's eating habits.

ADVICE FROM THE FARM

A Happy, Healthy Home

Our experts discuss how they keep their rabbit housing and residents hopping.

A Fresh Start

"Sanitation is the key to your rabbits' good health, so cleaning and maintaining is a daily commitment. It may take several arrangements before you find the setup most conducive to the maintenance of your herd. You cannot let droppings and urine build up or you risk leaving your animals open to a host of diseases. (A bonus to this task is that rabbit droppings make excellent fertilizer.) It is also important to do thorough cleanings monthly, as cobwebs will need to be cleared, and caked-on droppings will need to be hosed. With monthly rabbitry cleaning, water bottles should also be washed thoroughly in hot water."

—Sami Segale

Nutrition Is Key

"Proper feeding and nutrition are fundamental in providing your rabbits with the right start. I prefer a 16 percent alfalfa-based pellet fed every night. I also like to give a handful of oat hay once a week to keep their digestive systems happy and running smoothly. Generally, I feel that my rabbits don't need any treats or produce. But if I am going to feed them these on special occasions, such as when traveling long distances or during times of elevated stress, I prefer to provide them with carrots, apples, or bananas. I stay away from lettuce because of the high water content, as well as from many fruits that are too high in sugar. If using dried or processed treats, it is important to consider how the fruits or vegetables may have been processed before feeding them to your animals."

—Keelyn Hanlon, ARBA Judge

Comfort and Food

"I owe my rabbits' good health to keeping a clean environment and maintaining comfortable temperature control in the rabbitry as well as keeping them on a restricted diet. We use pine wood chips in the catch pans under each cage. Temperature is controlled in the rabbitry by the structure's insulation, fans (not trained directly on the rabbits) in the warmer months, and a dehumidifier in the winter months. Our rabbits enjoy a snack of timothy hay once a week and are on a strict diet of 2 to 3 ounces of Select Series PRO Formula Rabbit Food as a complete feed. We use Select Series SHO Supplement during show season. Rabbits can be put on this feed at least six months before the show. It improves skin and fur quality for showing."

—Brenda Haas

chapter **FOUR**

Rabbit Behavior and Handling

Rabbit keepers need to learn and practice proper handling techniques for many reasons. First, handling rabbits correctly is the only way to ensure the safety of both the rabbits and the keeper. Second, handling is the primary way that we teach rabbits that they can trust us. Mishandling, and therefore unintentionally frightening, the animals only fosters mistrust of human beings. Good handling practices engender calm and secure rabbits, in turn making them easier to handle. Last but certainly not least, all living creatures deserve our respect and kindness. Period.

Understanding where the rabbit is coming from in terms of instinctual behavior should be foremost when trying to catch it, take it out of its cage, turn it over, or carry it around. If you put yourself in your rabbit's place, you'll gain insight on how to handle it in a nonthreatening way.

How Rabbits Behave and Why

One thing you need to remember is that rabbits are at the bottom of the food chain. There's a reason they produce so many offspring—rabbits are dinner for many other animals, and reproducing in vast quantities ensures the survival of the species (even if many of those offspring become food for predators). Rabbits themselves are herbivores, solidifying their place on the chain—the only thing they hunt for is a good patch of grass.

In addition to rabbits' rapid reproduction, their eye placement also tells us that they're prey and not predators. A rabbit's eyes are set on the sides of its head, which allows for a nearly 360-degree view around it. This comes in as a handy defense when everything is trying to eat you. A predator's eyes are placed together at the front of the head, creating binocular vision. Because not many things hunt predators, they don't need to see all around them. Instead, they're able to focus in on their prey as it runs off.

Of predators and prey, guess which category you, a human, fall under? Exactly. You're a predator. You're a carnivore, or omnivore (generally speaking), and your eyes are set on the front of your face. I mention this because, although domestic rabbits have been around people for thousands of years, instinct tells them that it's still a strong possibility that you might eat them. Now all you have to do is convince your rabbit that you're one predator that isn't going to hunt it down.

This rabbit's ears, eyes, and nose are all attuned to its surroundings.

Natural Ability

Rabbits are incredibly aware of their surroundings and are ready to take off at a moment's notice should they sense that a predator is near. They have a highly developed sense of smell and excellent vision, even in low light. This makes sense considering that they're most active at dawn and dusk and, so, these keen senses come in handy. They're also tuned in to very low sounds—lower than we humans can hear. All of this goes to show how nature has programmed rabbits to be on high alert.

Rabbits prefer to keep all four paws on the ground at all times, which often results in frustration for the people trying to handle them. From the rabbit's point of view, a rabbit whose feet aren't on the ground isn't able to hop away and is probably about to become some predator's meal. It stands to reason, therefore, that when a rabbit hasn't been socialized with humans, it'll struggle, twist, and jump away from those trying to hold it. Other rabbit defenses include biting with their long teeth and scratching with their powerful hind legs and sharp toenails. In fact, when a predator has a wild rabbit on its back, this fluffy critter is actually able to literally tear open the attacker's abdomen.

Excavation Expertise

Domestic rabbits are programmed to dig—period. They'll dig out of enclosed

Did You Know?

Rabbits are often said to be nocturnal animals. Technically, they're *crepuscular* animals, meaning that they are most active and prefer to eat at dusk and dawn.

areas that don't have a bottom. They'll dig in flower beds, on carpeting, and on furniture. There's no point in trying to stop this behavior because they can't help it. Their ancestors needed this skill to dig warrens for protection and raising families. You just need to remember that rabbits will dig out of enclosed areas that don't have a bottom. The best way to keep this natural behavior in check is to supervise rabbits in your home and be sure that if you aren't supervising them in an outdoor run, there's a floor to the confined area. You can also encourage your rabbits to dig only in appropriate areas by providing them with a designated space to enjoy their excavating urges. A large terracotta flowerpot filled with soil works well, as does a child's sand box.

Another instinctual behavior that rabbits aren't going to give up—no matter how hard you try—is chewing. Rabbits' teeth grow continuously throughout their entire lives. Your rabbit is not chewing things to annoy you—it *needs* to chew on something. If you don't provide something for it to chew on, it'll find something on its own. This might end up being its wire cage (which can cause malocclusion) or perhaps your phone's cord. Baseboards, carpeting, electrical cords—they're all fair game. Chewing any of these things is extremely harmful both for the rabbit and for your home. Keep your bunny's teeth busy with grass hay, wood blocks, and apple branches to save you both some trouble.

What Rabbits Like... and Dislike

Now that you've read a bit about them, you should have a better idea about what rabbits might like and dislike. They certainly prefer to keep their feet on the ground, which is contrary to what we humans like to do with them. We need to handle rabbits as part of their daily care, but we also enjoy their company as pets and often want to

The instinct to burrow for safety is innate in all rabbits—wild and domestic.

Rabbits enjoy each other's company, but because they can't choose that company for themselves, you'll have to carefully choose for them.

hold them in our arms. In addition, if you exhibit your rabbits, they'll be picked up and placed on their backs throughout their lives. It stands to reason, then, that a gentle, consistent approach to these tasks is necessary in order for rabbits to become familiar with these practices and learn to trust us as handlers. Rabbits do seem to enjoy stroking and face rubbing, and they are most relaxed with a regular routine in the rabbitry.

As far as temperatures, rabbits seem to enjoy the same climates as humans or even a bit cooler. They not only dislike but also won't survive in extreme heat. It bears repeating that rabbits can take the cold extremely well as long as they have protection from the elements, but they can't tolerate high temperatures for very long.

In companionship, many rabbits enjoy sharing a cage with a friend. Remember, our domestic rabbits are related to European rabbits, which live in colonies. Unfortunately, providing companionship isn't always as easy as it sounds. For example, two intact bucks (male rabbits that are not neutered) won't make good cage mates. They'll fight for territory (and any does in the vicinity) and—prepare yourself—end up literally tearing off each other's testicles. If you'd like bucks to live together, they'll need to be neutered to create harmony. Does can often live together without being spayed, but some are aggressive and refuse to share cage space. You can try to combat this by placing them in a home together that did not belong to either doe previously. Obviously, you shouldn't keep an intact buck in the same cage as an intact doe unless you plan on breeding them. If you're keeping all of your animals intact for breeding or showing purposes, then the only roommate option you'll have is a pair of does that get along. The exceptions to these roommate situations are rabbits that just don't want to play nice and pregnant does. You do *not* want to house two does together when one (or both) has a litter. This situation can end in disaster if the mother doe becomes protective of her kits or if her roommate sees the kits as intruders.

While I do have a pair of does living together, I also have another extremely aggressive doe that has never gotten along with other rabbits. My solution is to arrange my rabbitry so that the rabbits are close in proximity to each other for company but are not actually sharing a cage.

Aggressive Behavior

While they may look cute and cuddly, rabbits are completely capable of showing aggression should the situation warrant it. They may growl, lunge, bite, and scratch with both their front and back feet.

If your rabbit hasn't displayed aggression in the past, your first step is analyzing why its personality has changed. A rabbit might act aggressively for several reasons, the most likely being that the animal is injured or uncomfortable. Give it a good physical once-over, keeping an eye open for fur or ear mites, lacerations, or soreness, especially in the leg or paw.

If your rabbit tends to randomly nip or lunge at you and you're sure that it's physically healthy, begin responding to bad behavior with a sharp yelp or "Ouch!" If you were attempting to give it a treat and it lunged because it wanted the treat faster than you were offering it, remove the treat. Don't reward the rabbit with the object of its desire or you'll reinforce the bad behavior. Conversely, try to use positive reinforcement whenever possible, giving your rabbit a treat or some special attention as a reward for good behavior.

Ears pinned back is a sure sign of an unhappy bunny. Beware!

Whatever you do, don't ever hit your rabbit for any reason. The honest truth is that the rabbit won't learn a thing from being hit other than to become more fearful and, therefore, more aggressive. If you have a rabbit that shows aggressive tendencies and you're not comfortable handling it, ask a rabbit-raiser that is more seasoned in handling aggressive rabbits to help you. Turning bad behavior around isn't easy if the rabbit makes you nervous.

Understanding aggressive behavior is the first step toward dealing with it. The following are some types of aggression and the possible reasons behind them.

The Alpha Rabbit and Aggression Toward People

Like all animals (humans included) rabbits living together in the wild instinctually develop a hierarchy within the colony. The alpha rabbit is the pushiest within the group. Dominant behavior allows it to be, for example, the first one in line for food. This is a natural behavior that a rabbit often extends toward its keeper. This, of course, won't sit well with you because you're not willing to be bitten or scratched. Try using positive reinforcement to coax the rabbit out of its aggressive behavior toward you over time.

Another reason a rabbit will show aggression toward humans is fear. Fear

Placing your fingers near a rabbit's mouth can invite a bite, whether or not the rabbit is generally aggressive.

can result from an experience with mishandling, abuse by someone, or being chased by people or pets. In addition, if the animal isn't socialized properly and has no experience dealing with people who want to pick it up and cuddle it, it may react fearfully to these types of advances. This is why it's so important to socialize your rabbits from a very young age. The best way to retrain an animal that's afraid is to handle it frequently for short periods of time.

There is one form of aggression, however, that you won't want to correct or minimize. Very often, you'll find that a normally friendly doe becomes dominant and protective when she has new kits. You can't really blame her for that. In fact, you can be proud that your doe is showing signs that she's a good mom.

Food Aggression

To be clear, "food aggression" could technically be called "human aggression," considering it's the human that is attacked when food is brought to the cage. However, the behavior is about the *food* as opposed to the *human*. So we'll talk about them as separate concepts under the same topic.

Rabbits that literally bite (and growl and lunge at) the hand that feeds them are acting instinctually. For a wild rabbit to survive, it has to be the first one to the food, especially in the winter when there may not be many green things to munch on. A rabbit that defends the food it finds is a rabbit that stays alive through the coldest months. *You* are aware that you're bringing the food to the rabbit (not stealing it). But to your rabbit, the action of putting your hand into the cage (your rabbit's domain) and quickly pulling it out in reaction to the lunge looks a lot like a rival rabbit going for the food and then retreating from the aggression. As a result, your rabbit feels it successfully defended the food it needs to survive.

Sounds Like a Rabbit

Rabbits have long been considered "a quiet pet." And it's true that they won't bother the neighbors. For the most part, rabbits communicate with people through their body language. Activities such as running, leaping, and flopping over onto their sides all point to a bunny doing the happy dance, for example. But, the longer you're around rabbits, the more likely you are to hear from them, too. Below is a list of sounds that you may not have realized rabbits make because many of them are made at a very low sound level:

Growl. A rabbit can certainly growl, and this usually precedes a lunge and possibly a bite. If the rabbit perceives that you are a threat, it'll have no qualms growling and lunging at you. Forewarned is forearmed.

Snort. Snorting can come before or along with growling.

Cluck. It's not exactly the same clucking sound a chicken makes; it's a lot quieter. The rabbits' version means that they're satisfied with what they're munching.

Purr. Purring for a rabbit is lot like purring for a cat in that they're both sounds that mean happy and content. However, cats purr using their throat while rabbits make the sound by lightly rubbing their teeth together. It's a very soft sound, but one you'll want to listen for.

Hiss. This is exactly what you think it is. The hiss is used to ward off other rabbits.

Hum. I think all rabbits do it on occasion, but most rabbit keepers associate it with an unaltered buck wooing his ladylove.

Whine or whimper. Rabbits will make this noise if they don't want to be handled, although I most often hear this from a pregnant doe that has been put into a cage with another rabbit (especially with a buck)—either she doesn't want a cagemate, or she doesn't want a buck making useless advances.

Grind teeth. A rabbit grinding its teeth is almost unmistakable. It's hard to confuse it with purring, even though it's made the same way. If your rabbit is grinding its teeth, it's in a lot of pain and needs medical attention.

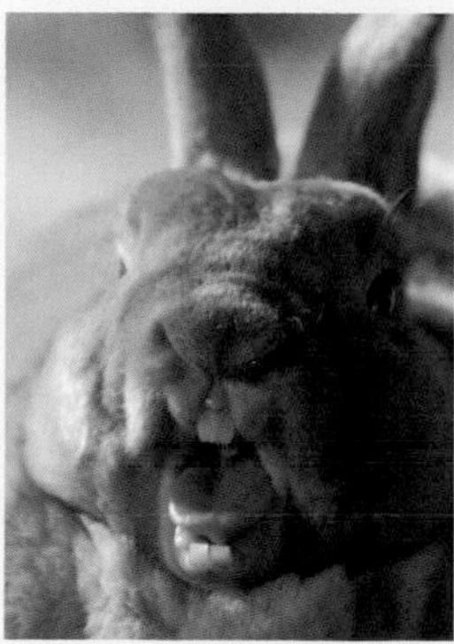

Scream. The sound of a rabbit screaming will send chills down your spine for two reasons. First, it sounds eerily close to a terrified child. Second, rabbits only scream when a predator is chasing them down or they're dying. It's never a false alarm when a rabbit screams.

Unfortunately, this may just be a part of your rabbit's personality that you'll have to live with—there's no bunny therapist to convince it that it doesn't need to defend anything. But you can try some of the following tricks.

Instead of using the food bowl exclusively, place multiple bowls of your rabbit's food in different spots around the cage. The idea is to get the rabbit to feel comfortable and realize that it has food all over the place and doesn't need to defend what you're bringing in. (Or, the extra food can serve as a distraction while you put another food bowl inside.)

If you prefer to feed from a single food bowl, try placing it into the cage in a different spot every time. Moving the bowl from a spot that the rabbit associates with defense may help it snap out of its aggressive routine. (Or, changing the bowl's position may just give you an extra second or two to dodge an attack.)

Hand-feeding your rabbit treats, such as carrots, may help it associate your hand with *bringing* food, which could lead it to be kinder to you during feeding time. You can use a long carrot to keep your hand out of striking distance, if you want.

If all else fails, you might consider using a J-feeder instead of a bowl or crock; your rabbit may still lunge, but your hand remains outside the cage, so there's no fear of getting bit.

Handling Rabbits

Rabbit keepers handle their rabbits for many reasons relating strictly to their care. But, quite frankly, if you like rabbits enough to raise them, you'll more than likely find them irresistible when it comes to just holding and cuddling them for fun. And if you're going to show or sell your rabbits, training them to be handled means that they'll be better behaved on the show table and in front of potential buyers.

If you've just acquired your rabbits (or you've just added a new rabbit to the rabbitry), let them become used to their surroundings for several days or more before handling them. To get them familiar with you, place them on a towel on your lap or on the floor and do some calm petting. Do this a couple of times before you begin handling them regularly.

Picking Up and Carrying Rabbits

Always remember to put yourself in your rabbit's position. Again, rabbits prefer all four feet on the ground at all times. Your job as the rabbit handler is to pick the rabbit up and hold it in a way that makes it feel as secure as possible. It may kick those strong hind legs at you; and remember, thick, long toenails are attached to those big feet. When those nails connect with your skin, it's incredibly painful. The smartest idea is to wear a long-sleeved shirt when handling

Begin handling your rabbits often (and gently) when they're very young.

a rabbit, at least until you're certain that it's no longer afraid of you.

The best way to practice lifting and carrying your rabbit is by using a large, flat table that's about waist-high. Place a towel or rug sample on it so the rabbit has something to grip when it's put back down. To pick up a rabbit, put one hand under it, just behind its front legs. The lower part of its chest will rest in your hand. Put your other hand underneath the rabbit's rump. The idea is to actually lift it with the hand that's under its front legs and chest while the hand on its rump supports its weight. Now, bring the rabbit against your body for extra support, with its head facing toward your elbow. This lift is similar to holding a football, and for that reason, it is called the football hold.

During the first session or two, the rabbit may get away from you and, in the process, leave a scratch. Although it's painful, try to remember that the animal isn't scratching you out of spite or aggression; it's truly afraid and wants to protect itself. The more often you handle your rabbit, the faster it'll relax and realize that it is a common practice without danger. Practice picking up the rabbit and using the football hold every day. Make these practice sessions short; picking the rabbit up a few times over the course of ten or fifteen minutes is plenty. After a couple of days of practice, walk with the rabbit around the table or a short distance away from it and back again. Then set your rabbit down and start over again.

Another way to pick up rabbits is by the extra skin on the back of their neck. Rather than putting your hand under the rabbit's front legs and chest, you grasp a fair amount of skin just behind the ears. Your other hand is once again placed under the rabbit's rump for additional support. However, with this technique, it's the rump hand that actually pushes up and does the lifting. The hand at the nape of the neck just secures and guides the lifting action. If you want to learn this handling style, it's best to learn it hands-on from someone who is experienced with the technique. Performing it poorly can hurt the rabbit, and no one wants that.

Derby is obviously comfortable enough in his environment to nap belly-up.

Whenever you're carrying your rabbit, make sure that you're able to safely put the rabbit down if it becomes frightened or begins to squirm too much. If you think your rabbit may fall out of your arms, simply drop to one knee. Your knee will offer a little more stability as you attempt to get the rabbit back under control; plus, it's easier to set the rabbit down in this position. If worse comes to worst, it's a much shorter distance for the rabbit to fall. If your rabbit jumps or falls from a long distance, it can break a tooth, a leg, or even its back. For this same reason, you don't ever want to leave a rabbit unattended on a table.

Turning Rabbits Over

This action may seem simple, but thanks to the rabbit's distaste for being belly-up, you do need to take your time teaching it to be turned over. Again, you're going to practice doing this using a waist-high table. When turning your rabbit, you'll use one hand to control its head and one hand to support its hindquarters. Flatten the rabbit's ears against its neck and back while reaching around the base of its head with your fingers. You should have a firm grip, without actually squeezing. A variation of this hold is to put your index finger between the rabbit's ears (close to the base of the head) and gently but firmly wrap your fingers around its head so that they point at its jaw. In a smooth motion, use the hand that's holding the rabbit's head to lift it up and slightly back, while at the same time tucking its rump toward you.

If the rabbit cooperates with you, the hand that was holding the hind end can now inspect the rabbit's body and teeth. But with rabbits that are new to being turned over, they'll more than likely wriggle free a couple of times. If this happens, try to gently guide them back to their normal position and try again. Don't attempt to force the rabbit into submission or you could break its back.

Training your rabbit to tolerate being turned over will benefit you later.

If the rabbit begins to struggle, use your free hand to grip the loin area; it may settle down. The real key to performing this technique correctly is learning how to have the proper grip on the rabbit's head, which prevents it from wrestling away from you. After you've practiced this hold successfully a few times, you'll be able to feel it when you have it right.

Another way to practice turning your rabbit over is to begin with the animal resting with its four paws on the front of your body. From this position, you can grasp its head and rump in the same way as you would from a table position, but now you can bend gently at the waist to let it rest on its back. This type of turnover doesn't feel quite so drastic to the rabbit, and it may feel more secure starting with this technique. However, if you plan on competing in showmanship in 4-H, you will have to teach your rabbit to let you turn it over from a table position.

Transporting Rabbits

You may need to bring your rabbit to a show, move it to another rabbitry, or just take it to the vet. Rabbit carriers are handy little cages made especially for rabbit transportation. They're usually made of the same wire mesh that wire hutches are made of and come complete with a detachable bottom tray. Carriers are designed to hold one to four rabbits. These mini cages are made to be just a little larger than the rabbit and with just enough space for the rabbit to turn around. Miniature, travel water bottles are usually attached to the fronts of the carrying cages. Rabbit carriers can be purchased wherever rabbit cages and hutches are sold, such as from online rabbit-supply companies, at rabbit shows, and from rabbit breeders.

A comfortable environment is just as important as food and water for any trek.

Rabbits may also be transported in small dog or cat carriers. If you choose to use one that isn't especially designed for rabbits, choose one made of hard plastic as opposed to one made of cloth so that the rabbit can't chew on it. Also, choose one with plenty of ventilation, as rabbits can easily become overheated. Cardboard boxes meant as temporary traveling carriers can be equally dangerous for a rabbit on a long ride. Think about your needs before choosing and purchasing a carrier.

Transporting by Car

If you're traveling in your car with rabbits, be sure they aren't sitting in direct sunlight, and never leave a rabbit unattended in the car. Rabbits are notorious for overheating. While en route, keep the temperature of the car comfortable for you—that should be fine for the rabbits. If the air conditioning is on, let it blow directly on the rabbits. If outside temperatures are extremely low, keep your rabbits cozy by adding bedding such as hay or straw to the carrier.

When traveling with rabbits, most people think to bring food and hay for them. I always suggest bringing extra water from home. This is easily overlooked, but often a rabbit will refuse to drink because the water offered at its destination has a different flavor or odor than the water it's used to. Generally speaking, rabbits won't drink while traveling in the car. However, once you reach your destination, they'll appreciate fresh water.

Transporting Rabbits by Plane

Transporting rabbits by plane is not uncommon, but you'll need to consider a few things before you head for the airport. Contact whatever airlines you're thinking of using and ask questions. Find out their rules about rabbits in the cargo area. Although most airlines do accept rabbits, some cargo areas are more comfortable than others. They may or may not have temperature control in the cargo area. If not, the airline may not let rabbits board the plane in extremely hot or cold weather.

If you're flying with only one rabbit or two, the airline may allow your rabbit(s) to travel in the cabin with you. Many airlines permit small dogs to travel in a dog carrier underneath the seat, so it doesn't hurt to check and see if the same offer is extended to traveling rabbits.

Whether you're shipping rabbits to someone or carrying them on board the plane with you, they'll need a United States Department of Agriculture (USDA) health certificate before stepping one paw on the plane. You can obtain a health certificate from a licensed veterinarian; it basically states that the rabbits are healthy and free from disease.

These certificates aren't expensive, and they're required by all airlines. Health certificates are time-sensitive documents that come with a deadline attached to them—usually around thirty days. Make the appointment with your vet just a couple of weeks ahead of your flight—don't go too early. Talk to someone at the airline you'll be using just to be certain of their requirements.

To fly, you'll also need a proper carrier, which is usually one of the all-wire rabbit carriers or a basic dog crate. The carrier is required to have your name and address on it as well as the recipient's name and address (if you're shipping the rabbits). The airlines expect you to provide food and water inside the carrier for the rabbits, and they will check that you did.

Permanent Long-Distance Moves

You can get your rabbits from one part of the country to another a few different ways. The first is to ship them in advance to another rabbit breeder in the new location so that they can care for them until you arrive. The advantage of this is that you won't have to worry about moving them *and* you (and possibly a family) all at the same time. The disadvantage is that the food (and water) will more than likely be different unless you ship some of each as well. The rabbits will also have to adjust to a whole new environment of sights, sounds, and smells without you. Then, after you arrive at your new home, they will have to transition all over again. You can also ship them directly to your new home if someone is there to pick them up and care for them until you arrive. This solution will at least keep the moves to a minimum. Don't forget that in this case you would need to travel ahead to set up rabbit cages at your new home.

Another option is to travel with your rabbits to the new home. If you're traveling by car, it's best to use a carrier that's twice the size of the rabbit to give it some comfortable wiggle room. Purchase an extra bag of rabbit feed to bring along with you. This will make it easier to transition them to a different feed should your brand not be available in that area (you'll combine a little of the old feed with the new feed until your rabbits get used to the new stuff).

This rabbit is being properly supported under its arms and hindquarters.

ADVICE FROM THE FARM

Rabbit Wrangling

Our experts talk about rabbit behavior and handling techniques.

Stimulating Company

"When designing your rabbits' homes, stimulation should be kept in mind. Rabbits are actually quite smart and can benefit from toys, ramps, cage proximity to other rabbits, the occasional treat, and regular interaction and handling. It is nice to design the rabbitry as a place you too can hang out and enjoy your rabbits' company.

Regular handling, especially when rabbits are young, will help tame them to your touch, which makes showing less stressful for them and more enjoyable for you. Daily 'rabbit pen time' can be a welcomed break from a busy household."

—Sami Segale

Putting in the Time

"The Netherland Dwarf (which is raised primarily for show) is known for its even temperament. This is because most of the breeders of the ND raise this breed in the 'cage' environment. This breed, like any other animal, needs to be raised with plenty of TLC (tender loving care). If you don't hold the bunny, then the animal won't want to be held. The same principles apply when training this breed for show. You must dedicate at least one hour each week to working with the rabbit so that the animal knows what is expected. I have noticed that once a doe has been bred, she does get a little 'nippy' if a buck is around. You need to pay attention to her behavior or just keep her away from the bucks if you don't want this aggression."

—Brenda H

Lesson Learned

"The sport of rabbit hopping can teach a child to properly care for a rabbit, relate to the importance of handling a rabbit safely, share in helping others work with their rabbits, and experience the independence of ownership by training their own rabbit to do a sport they both will enjoy."

—Linda Hoover

Well-Handled Care

"Handling is an important part of raising rabbits because it is the best way to keep a close eye on your herd. Your animals should be comfortable being handled by you so that tasks such as nail trimming, grooming, evaluation, and judging are safe for you and for the animal. It is just as important that you be confident and safe in your handling of the animals so that the animals will feel safe."

—Keelyn Hanlon, ARBA Judge

What's Up, Doc?

What's up, doc? As it turns out, not much. When it comes to rabbits and vet care, the two don't often meet. Rabbits are not difficult to care for, and they usually stay healthy. In fact, they have a lifespan of seven to ten years (longer than many marriages). Rabbits don't have any vaccination requirements, so routine vet visits are a nonissue. It's good to note, however, that veterinarians consider our domestic rabbits to be exotic animals. To be on the safe side, find a veterinarian in your area who specializes in exotics. Of course, if an animal shows signs of illness, such as lack of appetite, unusual lethargy, weepy eyes, a runny nose, or is in obvious distress (see sidebar "Signs of Pain or Illness" on page 83), then you need to visit a vet.

Often, rabbit breeders prefer to learn everything they can about their miniature livestock to avoid some or all of the costs of veterinary care. This may not be entirely possible, but most of the rabbit-raisers I know are well versed in rabbit ailments as well as their treatments. Whether you have just a couple of rabbits or a full rabbitry, it's prudent to learn simple, safe home vet care. If you have a large rabbitry, you can really save a bundle by knowing the basics.

Because rabbits are prey animals that sit at the bottom of the food chain, their natural instinct is to mask signs of illness as long as they can. To do otherwise in the wild would mean ending up as dinner. Domestic rabbits have the same instinct, which makes it easy to overlook their distress. This is one of the reasons many rabbits don't end up in the vet's office. By the time anyone realizes that something serious is going on, the rabbit is dead (or nearly so). This is why it's important not only to know some basic veterinary care but also to know your rabbits. If you handle your rabbits often, you'll be more likely to notice when something's off.

An Ounce of Prevention

All living creatures, no matter how well they're cared for, can contract an illness or have an accident. To minimize problems, it's important to provide the healthiest

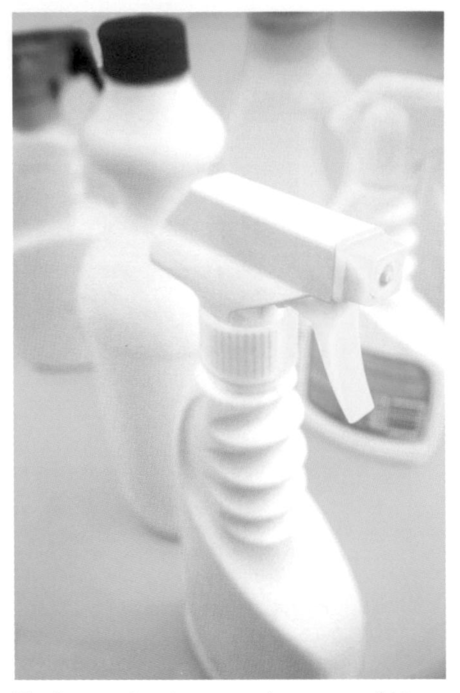

The best and easiest way to keep your rabbits healthy is to keep their environment clean.

and safest environment possible for your rabbitry. Most rabbit keepers have stories about averting common health issues by simply taking the time to prevent them.

Start with the most basic necessity—your rabbits' home. The best cages or hutches are spacious, solid, and secure against all predators. They protect rabbits from inclement weather but still have plenty of ventilation. Hutches should be thoroughly and routinely cleaned. This goes for any housing or structure that's built around the hutches as well. Proper housing includes the proper environment and, of course, the environment is ever changing. When it's warm outside, rabbits may need fans or misters around the rabbitry or frozen water bottles to lean against. When the temperature drops, you can offer them some sort of cubby lined with extra bedding such as straw. Like all other

Medication Warning

While some over-the-counter medications have been used on rabbits for decades (for example, sulfademethoxine) and may be perfectly safe to use in the appropriate doses, there are times when you'll need a stronger prescription medication. This means a trip to the vet's office. Of course, a good exotic-animal veterinarian would never prescribe something that's harmful to rabbits, but it's good to be aware of what's safe and what isn't in case a well-intentioned person suggests something he or she thinks will help. Something as simple as the wrong antibiotic could have adverse and even fatal consequences. The common oral antibiotics penicillin and amoxicillin are lethal to rabbits. Antibiotics that are safe for rabbits include enrofloxacin and sulfademethoxine. Never use another animal's leftover prescription medication without first talking to your vet. Most of the time, if an illness requires an antibiotic, you'll need to see a vet to obtain it.

Your Rabbit First-Aid Kit

The best way to avoid illness and disease is to start with healthy rabbits and use good husbandry practices in the rabbitry. That said, things can go wrong even under the best of circumstances. Having the right medications and supplies around will allow you to treat an ailing rabbit immediately, which will increase the likelihood of a full recovery. Although there are times when a call or trip to the vet may be necessary, the majority of minor rabbit illnesses can be treated at home.

Here are some things you should have in your rabbit first-aid kit:

- Antibiotic cream (the kind *without* the added pain relief, which can sting before it begins to help)
- Beneficial bacteria (probiotic)
- Corticosteroid or salve
- Cotton balls and swabs
- Hydrogen peroxide
- Ivermectin (or other wormer, such as fenbendazole)
- Miticide (or cooking oil)
- Nail clippers
- Papaya tablets
- Plastic medicine dropper
- Rubber gloves (disposable, if possible)
- Saline eyewash
- Small pair of sharp scissors
- Sterile cotton pads
- Styptic powder
- Sulfademethoxine
- Tetracycline ophthalmic ointment
- Towel
- Tweezers

Many medications are sold in larger amounts than you'll probably need (before they expire, anyway). One way of acquiring medications inexpensively is to buy them with another rabbit keeper and split the cost. I also like to keep an index card in the kit with the phone numbers of my vet and of the most knowledgeable rabbit-raiser I know (just in case the situation is more than I can handle). I can't tell you how many times a rabbit breeder has come to my rescue in an emergency.

If you get to know your rabbits well, you'll know when something's wrong.

warm-blooded animals, rabbits burn more energy in the winter in order to stay warm; give them some extra food so they have something to burn. If the hutches aren't housed in a barn-type unit, drop tarps around the entire rabbitry for cold weather protection. You should also provide extra nesting material for nest boxes in cold weather.

Timely hutch cleaning practices are paramount in maintaining the health of your rabbits. A reassuring extra step is to disinfect the cages from time to time to help prevent the spread of disease. You can use a bleach and water solution (1 part bleach to 5 parts water) with a wire brush on the cages—just be sure to rinse them well and let them dry completely in the sun. Many breeders disinfect a doe's cage before her litter makes its way out of the nest box.

You know that predators can be hazardous to your rabbits' health, but anything that makes your rabbits nervous—including the family dog—can also inflict damage in unexpected ways. Dogs shouldn't be allowed to roam loose around the rabbits, even if they're locked safely inside their hutches. Even dogs meandering calmly around the rabbitry can make a rabbit so nervous that it may race around its cage, flinging itself against the sides randomly. Sometimes a rabbit can do this hard enough to break its back, so it's worth your time and effort to make your rabbits' home as stress-free as possible.

Obviously, nutrition plays a big role in health. The number one thing that rabbits need is clean, fresh water and plenty of it. The right amount of fresh, quality food and a reliable feeding schedule will also go a long way in keeping rabbits issue-free. I offer not only fresh pellet feed but also fresh hay daily. If you're planning on switching feeds, you need to do it gradually or your rabbits may get diarrhea—sometimes literally overnight. As we know, rabbits become dehydrated from diarrhea very easily and it's often hard to turn the situation around (see the next page for more information). Diarrhea can be fatal, so proper nutrition and timing is extremely important. This means that, ideally, you should purchase the new food well before you run out of the current bag. Start by replacing just one-quarter of your rabbits' daily rations with the new pellets for a few days. Add a little more of the new feed every few days until it fully replaces the old feed. This is the perfect place to remind you that, thanks to rabbits' sensitive digestive

Signs of Pain or Illness

As prey animals, rabbits instinctually want to hide any pain symptoms or signs of distress to avoid unwanted attention by predators that will consider an ill rabbit an easy target. Below are common signs to watch for in the rabbitry. Not every symptom by itself means that a rabbit is ill, but each one is a red flag that tells you to look deeper into the situation.

- Lack of interest in food or water
- Consuming food slower than usual or dropping food when eating
- Teeth grinding
- Extreme weight loss for no apparent reason
- Unusually aggressive behavior
- Ragged-looking coat
- Manure caked on underside
- Sitting unusually (such as leaning back on its hind feet)
- Lethargy or looking depressed (you may notice that your rabbit's not curious about its surroundings anymore)
- Extremely soft droppings or diarrhea
- Difficulty moving (moving very slowly or not at all)
- Limping
- Head tilting (loss of balance)
- Discharge coming from its eyes or nose
- Dried and caked fur on the inside of its front paws
- Facing the corner of the cage, hiding, or sitting in a "hunched" position (if this is unusual behavior for the rabbit)
- Rapid breathing or signs of labored breathing
- Mouth wet with saliva
- Grunting or whining when moving or being handled
- Dullness in the eyes

systems, fresh greens have no place in young rabbits' diets. Adults may deal with *some* green treats just fine, but chances are that a baby won't.

Perhaps the best thing you can do for your rabbits' health is to observe and handle every one of them every day. This may sound time-consuming, but it'll go surprisingly fast after you get to know your rabbits and once your rabbits are comfortable with you—plus it's enormously beneficial and well worth the time. Handling your rabbits will let you feel each one's body and coat condition. Observing each rabbit for a couple of minutes every day will give you a lot of information about its normal habits. Know how your rabbits normally move, eat, drink, and behave when they see you. When you know each rabbit well, just a glance at the full dish of a food-loving rabbit will trigger warning bells. You'll be amazed at what early signals you'll catch and what disasters you'll avoid simply by spending a little time in your rabbitry.

Health Problems and Solutions

You can easily treat some of the following conditions, such as sore hocks and the first signs of diarrhea, at home. But others, such as hindquarter paralysis or wryneck, are going to need the attention of a veterinarian. If you're not sure what you're dealing with or you're uncertain of how to handle a situation, please seek the attention of a veterinarian. Some breeders feel comfortable handling the treatments discussed for these health problems, but this section shouldn't take the place of advice from your veterinarian.

Noninfectious Health Issues

Many noninfectious health problems are simple to cure. At the very least, animals with these conditions don't need to be quarantined as the disease can't spread to their furry neighbors. Here are some common noninfectious health issues and their respective treatments.

Corneal ulcer. Corneal ulcers are lesions on the eyes. The eye's cornea may look bluish, and you may notice that the rabbit is squinting or tearing. This usually results from a scratch on the eye or from very dry eyes. It's also possible that the ulcer may have developed a secondary bacterial infection. Temporarily place the rabbit in a dark area to lessen the irritation to the eye. You may want to consult a veterinarian about medication.

Diarrhea (nonspecific). Diarrhea is a symptom of a separate health problem; it's not a disease in and of itself. You'll want to rule out the possibility of a bacterial infection. If the rabbit seems

Good to Know about Oats

When I notice that one of my rabbits has watery or loose stools, I give it a big handful of oat hay or uncooked old-fashioned oats (not the sugary instant kind). Like magic, the rabbit's stools are back to normal. Note, however, that this trick only works if the rabbit doesn't have an underlying illness or condition causing the diarrhea.

fine otherwise, it could be that the flora in the intestines is simply off balance. Often, a rabbit being treated with antibiotics will have diarrhea. However, nearly every intestinal illness and parasite can cause it, as well. Diarrhea is a quick rabbit killer (rabbits become dehydrated rapidly); it requires immediate treatment. If the rabbit is currently on antibiotics, discontinue them until you can figure out the reason for the diarrhea. Add healthy bacteria to the rabbit's gastrointestinal tract by giving it a probiotic. Many rabbit breeders have found fresh oat hay (or old-fashioned oats) to be helpful for drying up diarrhea.

Fur chewing. If a rabbit is chewing the fur off of its body (or neighbors' coats), it may be bored or nervous or not have enough fiber in its diet. It could also be coming down with enteritis (inflammation of the intestines) or it could just be a behavioral issue. Try giving the rabbit in question some magnesium oxide and adding more fiber to its diet with hay or other high-fiber foods. Provide some toys for its cage to alleviate possible boredom, and keep it away from other rabbits until the problem is resolved.

Even something as simple as a particular kind of hay can resolve minor health issues.

Olaf's upset stomach can usually be settled with a little oat hay.

Heat stroke. Rabbits that are stretched out and breathing heavily may be suffering from heat stroke. Sometimes they will hold their heads high and their mouths open to aid in breathing. The fur around their mouth will be wet. If a rabbit in this condition isn't tended to immediately, death is almost certain. Rabbits are at risk for heat stroke when temperatures climb to 85 degrees Fahrenheit and above and possibly at lower temperatures in areas with high humidity. Watch out for quickly rising temperatures.

Affected rabbits should be relocated to a cool environment, such as an air-conditioned house or a cellar. They should remain there for as many hours as possible; certainly until it cools down outside. Try wiping the inside and

It's important to keep rabbits cool because they don't handle heat well. When it gets too hot for them, give them a frozen bottle of water to snuggle up to.

outside of the rabbits' ears with a cool, wet cloth or ice cubes wrapped in cloth. The ears are a rabbit's cooling system; blood vessels in the ears are closest to the skin, and cooling them will expedite cooling the rabbit.

If the rabbit is overheated to the point that it looks lifeless, every second makes a difference. Fill a container or sink with water that's at room temperature (*not* cold water as this could shock its system, and it could die from the shock). Place the rabbit's body gently into the water, supporting it the entire time. Don't let the rabbit's head go underwater—just soak it up to its neck. Afterward, take it to a shady, quiet place to recover.

Hindquarter paralysis. A rabbit that is dragging its hind legs around due to lack of motor control has dislocated his spinal vertebra. It could also be fractured as opposed to dislocated. This sort of trauma can occur if the rabbit gets excited or frightened all of a sudden and gives a big thrust with its back legs. It can also happen during tattooing if the handler doesn't learn proper handling techniques. Often, owners never find out what happened, but it's usually a grave prognosis. Most rabbit-raisers will have the rabbit *euthanized* to end its suffering. However, you should certainly seek immediate professional help for your rabbit.

Malocclusion. Also known as *buck* or *wolf teeth*, malocclusion occurs when the rabbit's top or bottom teeth grow extremely long and possibly crooked. When this condition is severe, the teeth interfere with the animal's ability to eat and cause wounds in its mouth. Some malocclusions can be addressed, and some can be fatal. It depends on the reason for the problem. If the malocclusion appears after a rabbit has been pulling on cage wire for a long period of time, try putting the rabbit into a wireless hutch to see if its teeth will grow back properly. Most of the time, however, malocclusion

Those Pearly Whites

Rabbits in the wild eat some particularly tough and abrasive plants, which quickly wear down their teeth. To remedy this situation, nature designed their teeth to grow about ½ to ¾ of an inch a month.

is a hereditary condition. If the elongated or crooked teeth are inherited, there isn't much you can do except clip the teeth regularly for the rest of the rabbit's life so it can eat properly. If the condition is so severe that the rabbit is constantly in pain due to sores or malnutrition, it may be best to have it euthanized. A rabbit with a malocclusion shouldn't be used as breeding stock. It's considered quite unethical to breed a rabbit with this health issue.

Slobbers. Excessive salivation will cause a rabbit to have a wet face, chin, or dewlap (the flap of skin under the chin of some rabbit breeds). The saliva may end up causing severe irritation of the chin, neck, and front legs. Slobbers is a symptom of another problem in the rabbit's body, such as an abscessed tooth or contaminated hay. Be sure that the rabbit isn't suffering from heat stroke, for which one of the symptoms is a wet mouth. A tooth problem (which is usually the cause of slobbers) requires treatment. If contaminated hay is the cause (it may be moldy upon close inspection or it may be soiled with urine and feces), remove and replace it with fresh hay. Some breeders prefer to feed a

If your rabbit pulls on its cage's wire, try putting it in a wireless hutch. This may seem like just an annoying habit, but gnawing on wire can really damage the rabbit's teeth.

rabbit with this symptom strictly pellets.

Sore hocks. This condition is when the fur on the bottom of the rabbit's feet wears away and sores or ulcers develop. It's most often seen on the hind feet, but it can happen on the front feet, as well. This is very painful, and the rabbit will look uncomfortable while sitting and try to rock back on its heels to avoid the sores touching the ground. The rabbit may also lose weight and look lethargic.

Sore hocks can be exacerbated by wire flooring, but contrary to popular opinion, it isn't the primary cause of the condition. Rabbits with a very thin layer of fur covering the bottoms of their feet, nervous animals (that thump a lot), and wet conditions on solid flooring can also cause sore hocks. (See the sidebar "Rabbit Feet and Welded Wire" on page 48 in chapter 3.)

To remedy the condition, place a solid, dry surface inside the cage for the rabbit to sit on. Treat the affected areas of the feet with astringent and follow up with a corticosteroid. Rabbits that are chronically affected by sore hocks might need to be placed permanently in a hutch with solid flooring. But be warned: if its cage isn't kept dry, the condition will worsen.

Hutch burn (urine burn). If the skin on the inside hind legs or genitals of a rabbit appears to have been scalded, it is likely a result of hutch burn caused by the rabbit's own urine. Sometimes the urine guards in the cage are at an angle that causes the urine to splash back onto the rabbit. Hutches that are dirty and wet can also cause the condition. A secondary infection can follow if the problem is not addressed. Correct the angle of the urine guards, clean and disinfect the cage,

If your rabbit has sensitive feet, pad its cage with cushy bedding such as straw or shavings.

You can probably imagine how easily an Angora's fur gets into its digestive system.

and use an antibiotic ointment on the affected areas of the legs and genitals.

Wool block (fur block). Wool block is a physical blockage (usually of fur and other debris) in the stomach or small intestine. Rabbits ingest wool and fur when they groom themselves or nibble on their neighbors' coats. The wool mixes with the undigested food in their stomachs or intestines and becomes a firm ball called a *trichobezoar*. Signs of a rabbit with wool block include lack or loss of appetite, weight loss, intermittent diarrhea, very small stools, manure strung together by fur, and a complete lack of stool. If any or all of these symptoms are present in a molting rabbit, it may have this ailment. When wool block is caught early and treated, the blockage can pass through the rabbit's system without a problem. But the key word here is *early*. The rabbit may not survive if the blockage is left untreated.

While wool block isn't limited to longhaired rabbits, the condition is predominately associated with the wool breeds. Rabbit-raisers with wooly animals should take preventative measures against this ailment. To help prevent a trichobezoar from forming:

- Make sure your rabbits' feed is high in fiber.
- Offer grass hay or oat hay for roughage as often as possible—preferably every day.
- Offer papaya enzymes one to two times a week in tablet form or whole form. (Papaya enzymes break down wool fiber.) Also try putting pineapple juice—which also breaks down fiber—in the water bottle.
- Keep longhaired rabbits groomed to eliminate as much loose wool as possible.
- Try not to overfeed a rabbit, as this could cause a digestive system backup and encourage wool block.
- Always have fresh water available for the rabbits to drink.

Rabbits can't vomit, so any blockage has to be pushed out the opposite end. If the blockage is severe, take your rabbit to a veterinarian. Surgery may be necessary. However, if symptoms have just appeared and your rabbit is still leaving some droppings, there are a few home remedies that you can try before taking a trip to the vet. Try to break up the blockage by treating it with one or two of the home remedies below. Try two home remedies for a day or two, then add another treatment. If it isn't working, bring your rabbit to the vet for more intense treatment. Don't wait too long because wool block can destroy the intestines quickly. "Too long" can vary between animals, but I'd say three days with treatment and without stool is a big red flag.

- A piece of banana (about one-third) with the skin still on.
- ½ to 1 teaspoon of stool softener on a banana if the rabbit will eat it. If it won't, you can use an eyedropper. If

using an eyedropper, put the rabbit on someone's lap with its feet resting on the person's legs. Try to insert the dropper at the side of the rabbit's mouth. Don't turn the rabbit over and force the medicine down its throat; it may aspirate the fluid into its lungs.

- A dropper or two of mineral oil per day over several days.
- 1 teaspoon of meat tenderizer mixed with mashed banana.

Wet dewlap. As mentioned, the dewlap is the big fold of skin just under the chin of a rabbit. Dewlaps are more pronounced in certain breeds than in others and are usually associated with does. Wet dewlap is indicated, obviously, by the fur on the dewlap becoming wet and matted. If this goes unnoticed, an infection can occur, and the skin will become irritated, turn green, and emit a foul odor. This condition is usually caused by the rabbit dragging its dewlap through a water crock while it's drinking. The nice thing is that you can easily cure wet dewlap: either replace the crock with a water bottle, or raise the crock up a couple of inches. Apply antiobiotic ointment to the irritated area (you may want to clip the fur off the wet area before applying the ointment).

Wryneck (torticollis or otitis media). You'll recognize wryneck right away—the rabbit carries its head tilted and often at an abrupt angle. This can accompany loss of balance or even falling over when the rabbit tries to hop. Wryneck is actually a symptom of inflammation of the middle ear. Usually, the inflammation is due to an upper respiratory infection or a bacterium called *Pasteurella sp.* You'll need the help of a veterinarian with this one. Your vet will run tests to discover what is causing the inflammation and will prescribe a medication to correct the problem. Wryneck isn't usually simple to cure. Sometimes the antibiotics work, sometimes they don't.

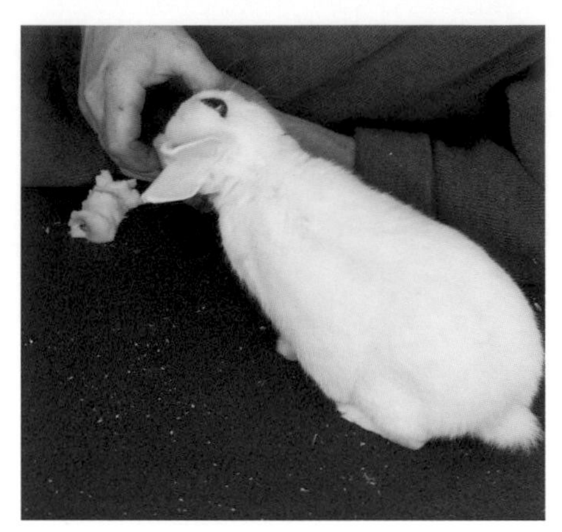

Petunia is suffering from wryneck and can't help but tilt her head as she grabs the apple.

Bacterial Diseases

The following diseases are infectious; that is, they're transmittable from rabbit to rabbit. Remember that the best defense is a good offense. In this case, adhering to excellent husbandry practices and rabbitry cleaning is a must. In addition, use a quarantine ritual for all newly acquired rabbits.

Abscesses. Although abscesses themselves aren't infectious, the reason the abscess is present in the first place might be. Rabbits can develop an abscess due to a scratch or bite from another rabbit, but abscesses can also pop up if the rabbits have the bacterium *Pasteurella sp.* in their bloodstream.

Abscesses show up initially as a lump somewhere on the rabbit's body, usually around the shoulders and head. Upon further inspection, you may notice that the skin over the bump is red and warm. Eventually, the abscess will burst and drain without any help from you. However, this leaves the wound open for even more bacteria to collect, and germ-filled pus from the abcess may spread to other parts of the rabbit. To prevent this, you should manually drain and treat the abscess before it bursts on its own. Small abscesses are easy to deal with, especially if you're proactive and drain the abscess yourself. (If you're uncomfortable doing this, then a rabbit breeder or vet is the next best bet.) Cleaning abscesses is easiest when you have another person to help you.

1. Have someone hold the rabbit so you can wipe medical disinfectant on the abscess bump. If you want to, clip some hair away from the area so that you can get a better view.

2. Take a brand new razor blade (or other sterilized sharp instrument) and make a slight cut in the center of the raised lump. It might seem like it hurts, but if you've ever had an abscess (or a pimple), then you know that the pressure from the trapped pus is actually relieved when it is expressed.

3. Use clean cotton pads or tissues to gently press on the sides of the abscess to push all of the pus out.

Knabbel is displaying his perfect, abscess- and sore-free feet.

4. Use the disinfectant to wash the wound (and the area around it).

5. Put an antibiotic ointment on the wound. (The human kind is fine, but don't use the type that says "pain reliever." The pain reliever in the ointment actually stings before it relieves any pain.) There's no need to bandage the wound—the rabbit would either pull it off or it would fall off on its own.

6. Wash your hands well with soap and water, and throw away the cotton pads or tissues used to clean the wound.

7. Apply some ointment every day to the abscess until it's completely healed.

Proper sanitation practices in the rabbitry will help prevent the infections that cause abscesses. Also watch that cage mates or breeding pairs don't fight, and be sure to separate littermates after weaning so that intact males don't start fighting over territory.

Foot abscesses. These usually look like little nodules under the skin on the toes, feet, or legs. Typically, they are caused by something directly irritating the foot. The nodules are abscesses and can become severely infected if harmful bacteria enters the site of the open wound. Soak the affected foot in an iodine solution several times a day. If the abscesses aren't draining by themselves, they may need to be lanced. Keep cages disinfected and surfaces smooth without anything protruding into the hutch that could pierce or snag the abscesses.

Conjunctivitis (weepy eye). Symptoms of conjunctivitis include redness or irritation of one or both eyes, possibly with some discharge on the lower eyelids and matted fur at the corners of the eyes. Weepy eye can be caused by bacteria and is often associated with an upper respiratory

Mucoid enteritis is often seen in young rabbits, possibly due to dietary change during weaning.

infection. On occasion, a blocked tear duct is responsible. Ophthalmic ointment applied two to four times a day for several days will clear up a bacterial eye infection, but if it's a blocked tear duct (which won't respond to the ointment), it'll have to be opened by a vet.

Eye infection (sore eyes). Sore eyes is a bacterial infection most often seen in kits still in the nest box. If the kits' eyes aren't opening around day ten, and they seem stuck shut, take a closer look. Take a warm washcloth and press it gently to the eye for a couple of minutes; the eye will loosen and pus will be underneath. With immediate treatment, the infection can clear up completely. But if the infection is severe, the rabbit could be blinded. Carefully flush the eye out with water, and apply some tetracycline ophthalmic ointment a couple of times a day. Thoroughly clean and sanitize the nest box, and be sure to practice excellent nest-box sanitation at all times.

Enteritis complex. Enteritis is inflammation of the intestinal tract, which can result in a pot-bellied rabbit with diarrhea. Infected animals are often found sitting with their feet in a water crock. Because death can often be the outcome of this disease, the best chance a rabbit with enteritis has is to see a vet. It needs broad-spectrum antibiotics, a high-fiber diet, and reduced stress. This condition requires aggressive treatment.

Mucoid enteritis. Refusing to eat, grinding teeth, and producing jellylike stools are all signs of mucoid enteritis. Another telltale sign is that the rabbit has a potbelly and, if you pick it up and *gently* shake it, you'll hear what sounds like water sloshing around inside its abdomen. This type of enteritis is often referred to as "weanling enteritis" because it's typically seen in young rabbits that are being weaned (five to eight weeks old). One theory states that the condition occurs during the transition from the mother's milk to pellets because of the changes in the flora in the intestines. Unfortunately, the prognosis is not good. Usually, in rabbits this young, they go from looking just fine to dying in twenty-four hours. Keeping rabbits on a high-fiber diet will generally help to avoid mucoid enteritis. For rabbits transitioning from their mother's milk to solid foods, have grass and oat hay in the hutch for them to eat as they come out of the nest box.

Pasteurellosis (snuffles). An upper respiratory infection, snuffles is often a precursor to other complications such as pneumonia. A rabbit with snuffles will sneeze and produce a white nasal discharge. Because no medication can effectively cure pasteurellosis, a rabbit-raiser should work diligently to prevent it. The affected rabbit may simply live with the snuffles symptoms showing up every so often until it dies, or it may die right away. That said, you can give affected rabbits antibiotics to prevent secondary infections. This isn't a bad idea, because it's often the secondary infection that does the most harm. Make sure the rabbitry has good ventilation and disinfect water bowls to help keep bacteria to a minimum. Be sure to isolate any rabbit with snuffles from the rest of the herd.

Pneumonia. A rabbit that is having trouble breathing and has bluish lips and a discolored tongue may have pneumonia. This is an infection of the lungs caused by a bacterium or virus, and it often follows on the heels of snuffles. Remember to quarantine any animal showing signs of pneumonia. The affected animal needs to see a vet immediately. The vet will treat the rabbit with a potent broad-spectrum antibiotic.

Tyzzer's disease. Tyzzer's disease, an infection of the bacteria *Clostridium spiroforme*, causes acute diarrhea and a rapid decline of health. This disease can kill a rabbit within three days. It also shows up most often in weanlings and is hard to distinguish from mucoid enteritis without a necropsy. Some rabbit owners have had success administering metronidazole to the affected rabbit when symptoms arise. You should isolate the sick rabbit immediately, and take it to the vet. Both rodent control (mice carry

Quarantine Practices

Quarantine practices in rabbitries are valuable for preventing the spread of disease. Forgoing the simple practice of isolating newcomers from the rest of the herd until you're certain they're disease-free can be a costly mistake. Quarantining is a free and uncomplicated practice that could save your entire herd. Simply house the new rabbit in an area that's well away from the main rabbitry for about four weeks. The amount of time is up to you; however, it can often take a couple of weeks for the rabbit to start showing symptoms—better safe than sorry! When you're on rabbit-feeding (or cage-cleaning) rounds, work on the newcomer's cage last. This way, if it does end up sick, you didn't carry its germs to the rest of the herd.

it) and good sanitation practices are key in preventing this illness.

Vent disease. Vent disease is actually the rabbit form of syphilis. It affects both bucks and does and its symptoms begin with inflamed, possibly scabby lesions on the rabbit's genitals. You may also find scabs on the affected rabbit's nose and mouth that ooze a yellowish discharge. Affected rabbits may refuse to breed, or a pregnant doe may miscarry her litter. Penicillin can be injected intramuscularly or applied topically on the lesions by a veterinarian. Be extremely careful about letting other breeders use your buck, and always check both bucks and does for lesions and scabs before breeding. Don't breed infected animals, and treat any animals that have been in contact with a sick animal.

Parasites

Parasites, by definition, are creatures that feed off of a living creature for their own survival. Rabbits have their fair share of critters that use them as hosts.

Coccidiosis. Coccidiosis shows up in two ways: the intestinal form and the hepatic or liver form. Both types show up with the same symptoms: diarrhea, trouble gaining weight, a potbelly, and poor condition of the rabbit's fur and flesh. Affected rabbits may also be prone to secondary infections. Rabbit breeders are using amprolium or sulfa-demethoxine to combat coccidia in their rabbit herds. Again, practicing good rabbitry sanitation is paramount.

Cuterebra fly (warbles). The Cuterebra fly, or bot fly, often creates lumps around a rabbit's neck and shoulders (but they're not necessarily limited to these areas). At first glance, these may appear to be abscesses, but upon closer inspection, you should see a breathing hole in the center of the inflamed lump. The fly has penetrated the rabbit's skin and laid its egg there to feed off of your rabbit. While it's growing, it continues to feast on your rabbit until, one day, a mature flying insect crawls out of the hole. It's obviously in your rabbit's best interests to remove the larva before too much feeding takes place.

Removing the fly larva is exactly the same as draining an abscess—with one main difference. You want to carefully remove the larva so that you don't smash or rupture the insect while it's in the rabbit. Doing so can send the rabbit into shock and may lead to its death. If you've never removed a warble before and don't have the help of an experienced person, a vet visit may be in order. Again, this is why you need to get a good handle on keeping flies down in the rabbitry.

Flystrike. Flystrike occurs when a rabbit has wet, loose stool or doesn't clean itself properly. Flies see wet feces and urine as an invitation to lay their eggs. The eggs are usually laid in the rabbit's genital area (check the side pouches next to the rabbit's sex organs). Once the eggs are laid, they quickly hatch into maggots and begin burrowing into the rabbit. They eat the rabbit's flesh as they grow and can make their way deep inside a rabbit's body. The rabbit then goes into shock from the maggot infestation and dies soon afterward. This whole process can take only twenty-four hours. Flystrike has nothing to do with the rabbits getting less than optimal care. It just takes one day of pasty stool and a single opportunistic fly for flystrike to happen. You'll need to remove all of the maggots from the affected area for the rabbit to recover. Hold the affected area under running water; this makes the maggots surface. Once they surface,

One thing rabbit keepers are constantly battling is the fly population. Because fly spray is not recommended for rabbits, look for innocuous traps such as this sticky tape.

use tweezers to remove every maggot. Be gentle because usually the area is swollen and sore. Afterwards, dry the area so that flies aren't attracted to the moisture. Keep your fly-stricken rabbit indoors until it's fully recovered.

This technique works well for me most of the time. If flystrike isn't caught right away, though, there isn't much you can do. If you can't bring yourself to get near the maggots, a trip to the vet is in order. If it's a bad strike, you should see your veterinarian anyway. A vet may also be able to give you a product that kills fly eggs before they have a chance to hatch. Note: never use a household fly repellant on the rabbit cages and certainly not on the rabbit itself!

Ear mites or ear canker. Symptoms of a rabbit with a mite infestation include head shaking and scratching, which can lead to cuts and bleeding where the rabbit has scratched. The rabbit may also have yellowish scabby material on the inside of the ears. Isolate rabbits infested with ear mites, and treat them for three or more days with either a mitacide or olive oil in the ear canal.

Encephalitozoon cuniculi. This parasite can be the cause of neurological problems in rabbits and can cause a wryneck situation just like *Pasteurella* (but this time the problems are caused by a parasite). It can also affect the spinal

A rabbit's ears are especially sensitive and help regulate its temperature. The last thing you or your rabbit want is an ear-related health issue such as mites.

cord, heart, kidneys, and brain. The parasite is sloughed off when the rabbit urinates and is transmitted from rabbit to rabbit via contact with an infected rabbit's urine. This makes it easy for a doe to pass it on to her kits. *E. cuniculi* is difficult to diagnose without a necropsy. Unfortunately, this means that by the time the symptoms appear, the internal damage has been done. There isn't a treatment for *E. cuniculi* at this time.

Fur mites. The symptoms of a fur-mite infestation are simple: noticeable fur loss on the head, neck, face, and base of the ears. Assuming this isn't due to a natural molt, isolate infested animals first. Depending on the severity of the infestation, a simple application of kitten-strength flea powder may do the trick. If the mites are persistent, you can give ivermectin orally or topically, or inject it subcutaneously.

Pinworms. Pinworms are internal parasites that won't show any clinical signs unless there's a major infestation. An infested rabbit is difficult to diagnose. Sometimes you can see the pinworms by the rabbit's anus or in the rabbit's droppings. You can also bring a stool sample in to a vet for positive identification. Piperazine is the favorite dewormer for pinworms in rabbits. You can use ivermectin, as well. However, ivermectin is the most valuable when treating for roundworms. Some rabbit keepers periodically give deworming medicine to the entire herd as a preventative. Rabbits need a very small amount of dewormer. Usually, a practical dose is about the size of a pea. When in doubt about dosages, seek the help of a veterinary professional.

ADVICE FROM THE FARM

The Healthy Rabbit

Our experts offer some sage advice on keeping rabbits healthy.

An Ounce of Prevention

"At-home rabbit healthcare should focus primarily around prevention rather than treatment. By practicing appropriate husbandry, most rabbit breeders will drastically reduce or completely eliminate incidence of disease. If a problem occurs in the rabbitry that involves a group of rabbits, the breeder should immediately look to their husbandry techniques first and troubleshoot all potential areas of concern."

—Jay Hreiz, VMD

It's the Little Things

"Rabbits are prey animals, so they hide illnesses well, which challenges their keeper to stay alert. Something as seemingly insignificant as your rabbit not coming to the front of the cage as it normally does when you approach may be a sign that your rabbit is just not feeling like itself."

—Sami Segale

A Little Something Extra

"Supplements that can be added to the rabbit's diet to help improve fur condition are cod liver oil, linseed oil, dry bread, sunflower seeds, and pumpkin seeds. To improve the body condition, try adding some calf manna, raisins, sweet corn, apples, apple twigs, comfrey, and sweet feed."

—Linda Hoover

Try This

"If a rabbit isn't eating, try simethicone (infant gas drops). If you have some weepy eye problems with a doe, try putting her in a hutch with another friendly doe. They clean each other's eyes, and this sometimes solves the problem."

—Betty Chu

chapter **SIX**

Breeding Like Rabbits

Many people who keep a rabbitry intend to raise rabbits for show, fiber, or meat. These activities usually involve enlarging and improving their herd through breeding. The most valuable thing you can offer your breeding program is a rabbitry full of healthy and happy rabbits. Neither overweight nor underweight animals are desirable for a breeding program meant to improve a rabbit line. The pair chosen to breed should be free from any abnormalities such as malocclusions or what are considered "faults" for their breed on the show table. Make your primary focus about health, and let the breeding begin!

Rabbit Reproduction

Does don't come into heat (estrus) in the same way that mammals such as domestic dogs and cats come into heat. Rabbits are *induced ovulators*. This means that they release an egg only *after* breeding with a buck. So, it's possible for a doe to accept a buck for breeding at any time of the year. However, there are certain times when does are more likely to accept a buck's advances. Look for signs of the doe's acting restless or rubbing her chin on the feeder and water-bottle nipples. If her genitals are inspected, the vulva of a doe that's ready to breed will be a deep reddish color (as opposed to a pale pink).

An interesting fact to note is that the uterus of a doe is Y shaped. They actually have two uteri (called *horns*)—both usually carry kits from breeding with a single buck. It is possible, however, for a doe to become pregnant by two different bucks in either horn. This makes the litters of each horn due on two different dates. Unfortunately, it is likely she will lose both litters due to malnourishment or other developmental issues.

Choosing Rabbits to Breed

The first thing you should think about when choosing a pair of rabbits to breed is how their physical characteristics complement each other. If you've never chosen a breeding pair before, it's in your best interest to seek guidance from a seasoned rabbit breeder. If possible, the breeder should raise the same breed as you so that he or she will know exactly what to look for as far as desirable traits in the breed.

Once you've chosen the buck and the doe candidates, you'll want to look them over for physical health issues. You're looking for signs of illness, disease, and malocclusions. You want clear-eyed rabbits at a healthy weight in your breeding program. Make sure both of the buck's testicles are descended. The breeding pair both need to be adult animals (see the sidebar "Breeding Ages") so that they've reached physical maturity—this is especially important for the doe that has to carry and nurture a litter.

The Best Time to Breed

The health and well-being of your rabbit and her kits depends not only on good timing but also on your own practical scheduling considerations. Does give birth approximately thirty-one days

Sexing Rabbits

How can you tell a buck from a doe? Determining males (bucks) from females (does) is called *sexing* and is easy when you know what to look for, especially with adult rabbits. Young rabbits can be trickier, but you get better at it with practice.

With one hand, hold the rabbit's head and the scruff of its neck in one hand while the other hand cradles the rabbit's rear end. Place the first and second finger of the hand holding the rear end on either side of the rabbit's tail. Use your thumb to gently press down in front of the rabbit's genitals. If the opening is round and you see a protrusion coming out of it, this is a buck. Of course, the other indication that you're holding a buck is that he'll have two testicles on either side of the area where you pressed down. If the opening is more of a slit with no protrusion (and of course, no testicles), the rabbit is a doe. If you're checking the sex on an adult rabbit, you'll see testicles on a buck right away; there's no need to press above the genitals to expose the penis. However, when sexing babies or youngsters, the testicles haven't dropped yet and you'll have to inspect further.

Start practicing with young kits at about three weeks old and double-check your findings a week or two later for accuracy. If you still aren't confident that you've properly sexed a litter, have a more experienced breeder sex them for you, and mark the inside of their ears with a B for buck or a D for doe using a fine-tipped permanent black marker. Have the breeder show you the sexing process as he or she does it. You'll be amazed at how quickly you'll pick it up.

If you're trying to perpetuate a specific breed, you'll want to make sure that your intact rabbits don't ever mix with other breeds.

after being bred, so factor this into your plans. Will you be available for the kindling (birth)? Will you need to get someone to cover for you? Or, if you're a beginner, will your mentor rabbit keeper be available to help you?

Weather is also a consideration. For those living in the coldest part of the country, does kindling in the middle of a snowstorm may not be optimal. Freezing conditions make it harder to save kits that are accidentally born *on the wire* (on the floor of the cage rather than in the nest box). It is also harder to keep the litter warm enough inside the nest box. Typically, a doe will pull enough fur

Breeding Ages

How old should a rabbit be before it's bred? The age of both bucks and does will depend on the general size of the breed: small, medium, or large. Don't forget that there's a difference between when a rabbit *should* be bred and when one *can* breed. Rabbits may be capable of breeding at three months old, but their bodies are still immature. The following are general guidelines for the optimal age to breed variously sized rabbits (with examples of breeds of those sizes).

- Small rabbit breeds (Netherland Dwarf, Dwarf Hotot): five months old
- Medium rabbit breeds (Dutch, Mini Lop): six months old
- Large rabbit breeds (French Lop, Flemish Giant): eight months old

Rebreeding

The most humane time to rebreed a doe that has just had a litter is when her kits are six weeks old. At six weeks, the kits will be weaned, and the doe will have some down time to recuperate before raising another litter.

from her abdomen to keep her babies warm, but on occasion, she'll fall short. This is just something to think about and to prepare for.

Finally, you'll want to think about how timing affects your rabbit's purpose. Some breeders choose not to breed a doe in the hottest part of summer because a juvenile rabbit's ears tend to grow longer in the heat (to help them keep cool). This may not seem important, but if your rabbit is destined for the show table, this could make its final ear length too long and disqualify it from competition. Rabbit-raisers interested in specific shows will also concentrate on how old the rabbits will be for competition. Many want the oldest junior animals competing in the junior classes; the older junior rabbits are more physically mature, which betters their chances of winning the class.

How to Breed Rabbits

For the most part, bucks and does are very capable of producing bunnies on their own. That said, rabbit breeders have found that a few pointers can help you avoid problems and get the does successfully bred.

Always take the doe to the buck's cage for breeding. Does are highly territorial,

One thing to look for in a breeding pair is the rabbits' ability to be in the same area without trying to injure each other.

but a buck is always willing to make an exception for a lady visitor. Some breeders are comfortable leaving the doe with the buck for a couple of days or a week to ensure successful breeding. I don't recommend this for a couple of reasons. First, if the doe isn't receptive to the buck's advances, you may not be around to break up a fight. Second, while you can still get a general idea of when to expect the kits, this approach is just too inexact for my taste. I would rather have a more precise date than "sometime within the next seven days."

If everything is as it should be—the doe is receptive and the buck is mature—the buck will chase the doe around the cage a couple of times before she'll let him mount her. If you've never watched rabbits mate, you're in for a quick education. The doe raises her hindquarters up, and the

Breeding Does in Tandem

One of the best pieces of advice about breeding I've ever received is to breed does in tandem. That is, never breed only one doe in the rabbitry if you can help it. I've tandem-bred every time, and it has saved many litters for me. Tandem breeding is especially important if a doe has never kindled and raised kits before. The idea is to breed at least one doe who has kindled (and successfully raised) a litter to weaning. If you don't have another doe that can be tandem-bred, ask another rabbit-raiser when he or she is breeding and if you can breed according to his or her schedule just in case. Here are some possible reasons to tandem-breed:

- A first-time mom might not have any idea what to do with the kits and might scatter them all over the cage. This doe usually shows no interest in the kits after that, and the kits are left to perish unless they are given a foster mom.
- A new mother may become frightened while kindling and start eating her litter of babies. As gruesome as it sounds, it's a common occurrence in nature. Try to rescue the remaining babies and have a more seasoned doe foster them.
- A doe may give birth to more kits than she can feed. A couple of the babies may begin to weaken from not getting as much milk as the others. In this case, having another mother foster a couple of the bigger, more robust babies is often the answer to saving the smallest ones.

It's worth mentioning that a clueless doe with her first litter may have no problem with subsequent ones. Give her a second chance, and decide from there if she's fit to be in your breeding program.

buck mounts her, breeds her, and then squeals and falls over. This all happens quickly, mind you, so don't take your eyes off them. Bring the doe back to her cage after she's bred. To be sure enough semen is deposited to create a litter, place her back into the buck's cage in a couple of hours.

Occasionally, the doe isn't ready to breed and won't raise her hindquarters for the buck. Instead, she'll turn and attack him. In my experience, this is a quick attack-and-retreat situation. Once the doe and buck pull apart from an attack, take the doe out before she decides to attack the buck again. There are teeth involved, so thick gloves come in handy for just such a situation. You can bring the doe back to the buck's cage in a day or two to try again.

Even more occasionally, the buck won't be interested in breeding. If your stud just sits there dazed and confused, you can try placing him on the doe's back. It's amazing how quickly they catch on with this simple maneuver. If he still is uninterested, you can return the doe to her cage and let her scent hang around his place for a bit. Then, bring her back over for another rendezvous, and he'll more than likely have figured out his first mistake. If for some reason, he still won't breed, it may be time to have her visit another buck.

Keep Good Records

Whether you're breeding rabbits for profit or for show, I can't emphasize enough the importance of good record keeping. Especially when breeding, no other practice will keep you more informed and on top of what's happening in your rabbitry as good notes will. Immediately write the breeding date on the hutch card as well as in your indoor records. At the very least, write the date down on *something*. Going back later and guessing is risky: you could easily get the kindling date wrong, resulting in the potential loss of kits. Some breeders do very well with a simple hanging calendar in the rabbitry. I prefer more detailed record keeping, but you'll figure out what works best for you.

Is She Pregnant?

There's a rather nonscientific method of testing a doe to see if she has *taken* (gotten pregnant). About ten days or so after the breeding date, place the doe back into the buck's cage. Usually, if she has taken, she will run around avoiding the male, possibly grunting or squealing. While this test can be helpful, some does will allow breeding again anyway, so this isn't a guarantee.

You can also weigh her on the breeding date, and then weigh her again in ten days. Assuming you're feeding her the same rations as before the breeding, if she gains somewhere close to a pound, it's another good indication that she might be pregnant. A more reliable test is to palpate the doe two weeks after exposing her to the buck and see if you can feel the marblelike embryos in her belly. Your hand should be just in front of her groin. Gently feel around the sides of the belly. Once she's far along, you can actually feel movement.

Preparing for Kindling

Of course, Mother Nature will primarily play birthing coach when it comes time for the doe to kindle. Still, we rabbit-raisers can do our part in preparing for the pitter-patter of fluffy feet.

Caring for the Pregnant Doe

During pregnancy, does don't require much in the way of special care. In fact, most breeders will suggest that you don't even increase the doe's feed until the second half of the pregnancy (day 16). Even then, breeders usually offer just a slight increase because obese does can have a harder time with pregnancy and giving birth. Pregnant does tend to drink more water, so be diligent about keeping plenty available. If you have an automatic watering system in your rabbitry, be certain to check the nipple in her cage daily.

The All-Important Nest Box

Nest boxes are extremely important for the kits survival. Our domestic rabbits' descendants burrowed into the ground and created a special warren in which to deliver and nurse their kits. A domestic rabbit is deprived of this, and instead we offer her a nest box with shavings at the bottom and some straw packed inside. I offer nest-building materials freely—the doe will get busy making her nest box just right. Place handfuls of nesting materials such as straw or oat hay inside the cage so that the doe can build her nest inside the box the way she likes.

When the doe has finished constructing her nest, it will be tunnel-shaped (this is her natural instinct). After utilizing the nesting materials, she will pull fur from her belly, sides, neck, and under her chin to make the nest not only cozy but also extremely warm. After the doe kindles, she'll pull more fur and place it over her kits.

The question of when to put the nest box into the doe's cage is a bit of a tricky one. Some breeders say to wait until a day or two before her kindling date so that the doe doesn't soil the box before kindling. Others say to put it in a little sooner just to be on the safe side. I fall into the second school of thought—better safe than sorry. Besides, only some does like to make their nest days early; most build it just hours before the kits arrive. Be sure to place the nest box inside the cage in the back corner opposite where the rabbit leaves most of her droppings. You definitely don't want the doe's box over her potty area.

Kindling

A quiet environment is best for a pregnant doe that's days from giving birth. Does preparing to kindle are focused on nest building, and their bodies are preparing

Hair from the doe's body helps create a warm nest for her hairless newborn kits.

About Nest Boxes

All rabbit breeders have a favorite type of nest box. Some like the ones made entirely of wood; they feel that it keeps the bunnies warmer. All-wood nest boxes should be carefully sanitized between litters using a bleach and water solution (1 part bleach to 5 parts water). Other breeders prefer the galvanized-steel type that has a sliding plywood bottom. The upside to this type is that it's easy to clean and easier to disinfect than solid wood. Many nest boxes include a small shelf toward the top of the box. This comes in handy when a litter is old enough to chase their mother around. Does will often jump onto the shelf and take a well-deserved break from their rowdy youngsters.

Another popular type of nest box looks like a wire mesh basket. The bottom and sides can be lined with cardboard, which can be discarded as it becomes soiled and changed to prepare for a different litter. These nest boxes are more often used in mild weather, as they're obviously not as warm as the other types.

Look around at as many varieties of nest box as you can find and, more importantly, ask other breeders which types they prefer and why. You can purchase nest boxes at rabbit shows or from rabbit-supply companies, and some breeders make their own. Nest-box size is determined by the size of the rabbit. For large rabbit breeds, the box should be about 20 inches long, 12 inches wide, and 10 inches high. For the medium breeds, the box should be 18 inches long, 10 inches wide, and 8 inches high. For the small breeds, the box should be 14 inches long, 8 inches wide, and 7 inches high.

This temporary nest box gives the rabbit keeper time and space to clean the real one.

Pregnant doe Bess has pulled hair from her body to pad the nest and is anxiously awaiting the arrival of her kits.

for labor, so the less commotion, the better. This is the case especially with does that are kindling for the first time. As previously mentioned, an anxious or fearful doe may resort to eating her new young or refusing to feed them.

Pregnant does don't generally make a big production when they're about to give birth, but they do show some signs that will clue you in to the coming event. If you have a good idea what the doe's eating habits are, the first thing you may notice is that she will eat less or go off her feed entirely a day or so before the kits arrive. The second clue is that she'll hop around gathering nesting material (that you provided) in her mouth to create a cozy nest. She usually does this on the same day that she kindles. However, the time frame can vary: she could begin to build her nest the day before she kindles or even a couple of days before—it will depend on the doe. As mentioned earlier, she'll also begin to burrow in the straw bedding and pull fur from her body and line the nest with it. Not only does the fur help keep her kits warm, the missing fur on her abdomen allows for easier access to her nipples. After kindling, the doe uses more pulled fur to cover her new babies like a blanket.

Does can kindle at any time, but most seem to favor the quiet, early morning hours. Often, the babies will have already arrived before you visit the doe's hutch in the morning. Sometimes a doe will kindle a little later, and you'll catch her in the act. The safest course of action is to stay back so that you don't disturb her or interfere with the birth. Although does give birth around day thirty-one, it could be a little earlier or a little later. About twenty-eight days after breeding, approach the hutch quietly when the doe isn't sitting in her nest box and peer into it—don't use your hands at this time. Watch for a minute or so and see if the fur she's placed inside the nest is moving up and down. If so, you can be sure she's already kindled.

Although does usually handle kindling quite nicely on their own (especially if they've raised a litter before), a couple of instances may require you to intervene in the birth. The first is if the kits have been born on the wire. If you catch them while they're still warm, simply place them deeply into fur inside their nest box. If they're cold, you'll need to warm them up before putting them into the next box. (See the sidebar "Warming Up Cold Bunnies" on page 111.)

Does and Kits

You've successfully bred a well-matched rabbit pair, you've taken great care of your pregnant doe, and now there's a brand-spanking new litter of kits in your rabbitry. Here's what you can do to help your doe-mom raise a happy and healthy litter.

Caring for Nursing Does

Veteran rabbit breeders have their own home concoctions for nursing does. These remedies are said to help anything from halting mastitis (an infection of the mammary gland that can be caused by streptococci or another bacterium in nursing does) to discouraging a nervous doe from eating her kits. A favorite remedy for does that have just given birth is to soak bread in warm milk and give it to the doe right after kindling. I have always given my new moms a piece milk-soaked bread, but more as a nice warm treat after her ordeal rather than as a magical cure. Most rabbit-raisers allow a doe to have as much pellet feed as she will eat after kindling as well as offer her daily nutritional snacks, such as small pieces of carrots, apples, and the like. Don't overdo the added snacks.

Caring for Kits

Rabbits have been kindling (giving birth) since long before people (breeders) ever came into the picture. However, that was before they became so domesticated and dependent on our care. It was also before people were concerned that kits stay alive. In the wild, many, many kits don't survive and many, many do. When we keep rabbits in cages, we place them in a setting

Word to the Wise: Angora Wool

If an Angora doe is in full coat when she pulls fur for her kits, it often comes out in long strands. These fibers often wrap around the delicate neck and feet of the newborns, which can be dangerous for their circulation and possibly life-threatening. Before your Angora kindles, you can trim her coat short (about the length of a puppy's coat) so that when she pulls fur, the strands are short. You can also gather the fur she's already lined her nest with, cut it into short pieces, and place it back into the nest box.

that doesn't necessarily allow nature to take its course. More bunnies survive in captivity, and it's hard for does that are trying to wean their kits to get away from them! So, we've placed ourselves in the position of becoming caregiver, nurse-maid, weaning attendant, and sometimes foster parent to our rabbit herds.

In the Early Days

Does that have raised litters before usually have this baby thing down on their own. That said, many breeders like to take a peek into the nest box for several reasons. First, it's nice to know how many kits the doe will have to nurse. If the doe has eleven babies in there and you have another doe with only a few, you may want to give a couple to the doe that has fewer so that all of the babies have the best chance at reaching a nipple to nurse. Second, if there are any stillborn babies, they should be removed from the nest box. Third, if the doe is inexperienced or has had a history of not caring for her young, then the breeder will want to take a look at the kit's tummies each day to be sure they're full and round, which means they're getting plenty of milk. Another reason, purely curiosity related, is that breeders just love to check out the possible coat colors. If you've never handled newborn kits before, there are ways to go about it and some things to take into consideration.

Personally, I like to distract the doe before I go near the nest box—not necessarily because the doe may be protective and therefore aggressive (although she might), but because I want her to see me as non-threatening. I feel that bringing my doe a goodie and letting her begin to munch on it before I remove the nest box to inspect it conveys my message of peace. You can offer her an apple slice, a carrot, or another favorite treat. Then remove the nest box so you can check on the brand-new arrivals.

Here is a litter of kits at one day old (top) and ten days old (bottom).

There are two schools of thought on handling kits from birth. One is that the doe may become stressed due to the human scent on her kits and kill them or refuse to care for them. The other is that if the doe knows you well, she'll have no problem accepting you touching them. Of course, handling the kits will make it easier to count and inspect them. If the doe is unfamiliar with you, or she hasn't been in your rabbitry for very long (she may have come to you bred by someone else), but you feel that you need to

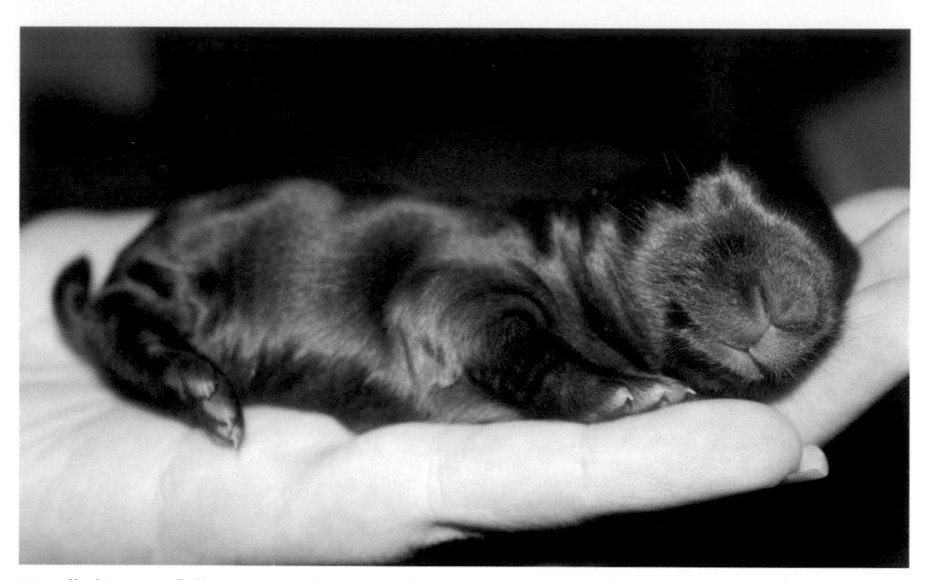

Handle kits carefully—especially where Mama is concerned—from the very beginning to ensure that they grow up accustomed to people.

handle the kits, you can put some vanilla extract (from the kitchen) on your finger and wipe it onto the doe's nose. This way the kits will all smell the same as far as she's concerned. If you want to err on the side of caution, gently push the fur and nesting material aside with the back of a wooden spoon to get a good view. Count the babies and be sure that your doe has enough nipples to feed them. Remove any placentas, dead kits, or extra bloody fur. Afterbirth and dead babies will attract flies, ants, and who knows what else.

Dead kits are quite common, especially in a doe's first litter. There are numerous reasons that babies are stillborn. Rather than focus on kits that didn't survive, try to focus on the babies that are healthy and alive because they're the ones that need you now. Dwarf breeds sometimes give birth to what are referred to as *peanuts*. These are kits that are half the size of their brothers and sisters and very often aren't fully developed. Some breeders say that they won't make it and pull them out of the nest box immediately. I leave them there with their siblings. Many simply don't make it, but I've had my share survive, too—especially if I give the wee one to a doe that has only a couple of kits in her litter.

Foster Care

Sometimes a doe will give birth to an extremely large litter, or will refuse to care for or feed her kits. Here's where you will congratulate yourself if you had the foresight to breed more than one doe at a time. While it's possible to hand-raise kits, it usually doesn't work out very well. Baby rabbits are extremely difficult to hand-feed and often end up aspirating the substitute food into their lungs, which results in pneumonia. The best thing to do for orphaned kits is to give them to another doe to raise. Simply remove them (or a few of them, in the case of an enormous litter) from their original box, and place them into the foster mom's nest

Warming Up Cold Bunnies

Once in a while, a doe will kindle on the wire, or a baby will be attached to a nipple and let go once its mom has hopped out of the nest box. If the kit is basically warm but has wandered, place it back into the nest with its siblings and it'll warm up quickly. You should check on does that have recently given birth periodically throughout the day just in case a baby ends up on the wire. Kits are born blind, deaf, and very naked. They become chilled quickly, and the exposure can kill them in minutes.

That said, don't ever immediately assume that a baby on the wire is dead. Kits will become very still when their temperature drops; this preserves their energy, which keeps them alive for the longest time possible. How much time has everything to do with how old it is, the temperature outside, and how long it's been on the wire.

There are a couple of different techniques to warm up a baby rabbit and save him. The first thing I do when I find what my husband affectionately refers to as a "Popsicle bunny" is to put it under my shirt against my skin. This begins to warm it up on the spot, and I often begin to feel the motions from its feet moving even before I get to the house. You can also turn a heating pad on low and wrap a towel around it. Put the heating pad and towel with the bunny inside a shoe box so that the heat surrounds it. It's better to warm the bunny up slowly. In fact, don't give in to the urge to turn the pad on high; the kit's skin is less than paper-thin and it can burn easily.

Another technique is to put warm water in a bowl on the counter. Now place the baby into a storage baggie (leave the top open and hold it securely), and place the storage baggie so that the baby is lying in the warm water. The idea is not to let the water touch the kit but to let the warm water heat its skin through the plastic bag. Yet another method is to put a lot of hand towels into the dryer, and when they're warmed up (not burning!), hold the kit in your hands with the towel wrapped around him. When the first hand towel cools, wrap a new warm towel from the dryer around him. Do this until the bunny pinks up and is quite warm without the towels, then put it back in the nest with its siblings. Before returning a Popsicle bunny to the nest box with the rest of the litter, it really needs to have no chill left in it. If it's still cool, the other bunnies will wriggle away from it; without their body warmth, it could become chilled again and perish.

Fair warning: Sometimes a kit can be brought back from the brink only to gape and gasp and not pull through, although it seemed warmed up. You've done your best, but some kits are just too far gone to be brought back completely. This is the not-so-fun part of rabbit raising.

box. If you feel more comfortable, use the vanilla-extract-on-the-nose trick so the foster mom doesn't realize that you've added more mouths to feed.

In a foster situation, don't forget that you want to remember which kit belongs to which mom. You're lucky if the markings of the original litter are completely different from those of the foster babies. If they're the same, mark the fostered babies ears with a permanent marker *before* you place them into the new nest box. In addition, don't assume that the mark will stay on the ear. More often than not, you'll have to mark the babies' ears every couple of days.

Continuing Care

Raising baby rabbits is best left in the capable hands of the mom-doe, barring any unusual complications. It's important to note that rabbits only feed their babies once or twice a day. The doe also sits over the babies for a couple of minutes a day and cleans their bottoms (which encourages them to eliminate) while she's in the nest box. Other than these times, the kits pretty much stay in their fur-lined nest box and mostly just sleep and grow.

The most important thing you can do for your kits is to keep the cage and nest box clean. After the doe has been visiting the nest box for a week or so, she may soil it, which is unhealthy for the babies (especially if they become wet from urine). Find a small container (just large enough to hold the kits comfortably), and gather up all the clean fur you find in the nest. Carefully take the kits out of the nest box, and put them onto the fur in the temporary container. You can now clean out the entire nest box.

Empty all of the soiled or damp materials and add fresh straw. Form the bedding into a nest shape and make a hole in the fresh straw. Now line the hole with the clean fur you saved. Add the babies to the fur-lined nest (try to save a little fur to place over the babies). You can even gently pull some fur from the doe's belly to add to the nest. Chances are, once the nest box is placed back into her hutch, she'll add some herself.

When the kits are about ten days old, their eyes will begin to open. They may open one at a time; that's fine. But watch out for any babies whose eyes aren't opening around the same time as or just after their siblings' eyes. If it seems to you that a bunny's eyes should be open and they aren't (within a couple of days of all the others' eyes), take a cotton ball moistened with warm water and gently run it on the eyelids to see if that helps coax them open. If not, the kit may have an eye infection. This is common with kits, which is one reason why it's important to keep the nest box as clean as possible. If you notice that one kit is having an eye problem, be sure to check the eyes of the entire litter. If the eyes are glued shut with a crusty material, there's a good chance that it is a sign of infection.

Eye infections can be simple to cure if caught in the early stages. But if they aren't dealt with immediately, they can result in permanent blindness. Use the moistened cotton ball on the crusty eye until the eyelid is clean. After that, use clean hands to gently pry the eyelids apart so that you can see the eye. Sometimes this simple procedure clears up the problem. If the rim of the eyelid is red, put a couple of drops of eyewash in the eye. Check the eye again in several hours and again the next day. If, at that point (or at any point), you see pus coming from the eye, then you'll need to put tetracycline ophthalmic ointment into the eye to battle the infection.

Treat an infected eye with the wash-and-ointment treatment for several days, and you should notice some positive changes. Once the eye is healthy again, it may be good as new, or there may be some cloudiness over the cornea of the eye. Cloudiness means that the rabbit is either partially or totally blind in that eye.

As the litter grows, you'll certainly notice that the hutch needs more cleaning than when it housed just the doe. It bears repeating that if you keep up high sanitation standards, you'll avoid a large number of health issues.

Weaning Kits

The first step in weaning kits from the doe is to remove the nest box. When the kits are about sixteen days old (give or take), they'll begin to hop in and out of the nest box where they've spent their entire lives until now. It's only fair to warn you that this is when beginning rabbit keepers often get sucked into the rabbit-raising world for life. There's literally nothing cuter than kits between the ages of two and four weeks old. Nothing comes close—not puppies, not kittens.

If you've been giving the doe greens, stop once the kits pop out of the nest box. You don't want the youngsters' digestive systems to deal with greens and treats yet. You should, however, have plenty of oat hay in the hutch, so the kits' digestive systems can transition smoothly from their mother's milk to pellet feed.

Typically, the nest box is removed from the hutch when the kits are three weeks old. If you live in an area with extremely cold weather, you may want to wait an extra week or two. But remember that the kits will now begin eliminating in the box, and this will make them wet, not to mention encourage bacteria to grow that can result in illness. So, remember to clean the box out regularly. A nice layer of straw as bedding on the cage floor will help keep the young ones warm, too.

Another reason to leave a nest box in for longer than the kits need it is actually to help the doe. The youngsters will attempt to nurse all day and will, in general, pester

When kits begin to venture out from their nest box, they're ready to be weaned.

As kits grow older and larger, you'll want to separate them from their mom and each other.

her to no end. If the nest box is the type with a small shelf over half of the top, she can escape them for a time.

Kits at three weeks old are ready to be handled by their caregivers. Handling these youngsters every day will make them easy to handle their whole lives and will get them on the path to becoming good companions and show rabbits. Young rabbits are fragile creatures that are injured and frightened easily, and their moves are fast and sudden. You can show older children how to handle kits properly, but kids younger than ten should only be allowed to pet them while someone older holds them.

Somewhere between week five and week eight, the kits should be weaned from the doe and separated from each other. The general rule of thumb here is that the smallest breeds are weaned closer to the five-week mark and the largest breeds at the eight-week mark. From a human perspective, this may seem sad; but from the doe's perspective, I assure you, it isn't. Her body works very hard to produce milk for kits of this age and she'll be grateful for the relief that weaning brings.

Some breeders wean the kits from the doe all at once, but I always remove half of the litter at a time. This allows the doe's milk production to reduce gradually and lessens the chance of engorgement in the doe's teats. You can take half of the litter out of the doe's cage in groups organized by sex or by number. Seasoned breeders tend to sex rabbits easily at this age, but it does take practice (see the sidebar "Sexing Rabbits" on page 100).

If you separate them by sex, be sure to double-check your conclusions a week or two later. Young rabbits have been known to "change sexes" due to an earlier misreading of their sex. The first group can be housed together in a temporary cage, or they can be placed into their own cages at weaning time. A few days later, you should separate the second half from the doe as well.

Many breeders will tell you that the youngsters can be caged together up to four months old. My advice is to resist the urge to house littermates together for more than a couple of days after weaning—otherwise, days tend to lead to weeks, and weeks lead to does being bred by their brothers, and bucks fighting to the point of removing (or attempting to remove) each other's testicles. Rabbits grow up fast, and time has a way of slipping by.

If you'd like to bide your time as far as having to provide cage space, you can always keep the doe kits with their mom for a bit longer. At about three to four months old, the young does may start to get territorial, but this arrangement usually works for a while.

Pedigrees, Tattooing, and Culling

Weaning time is the perfect time to do other important tasks for the new kits, such as filling out pedigrees, tattooing, and culling. As far as pedigrees, you should fill them out for all of your stock. Yes, some may go to pet homes where papers are unnecessary, but most people still enjoy seeing where their rabbit came from and keeping track of their rabbit's age. Pedigrees also have your rabbitry information on them, which is a handy way to be certain that potential buyers can reach you if they have any questions or if they want to share your name with someone else who wants to purchase a rabbit. And remember that pedigrees are mandatory if you plan on selling purebred show or breeding animals. In any case, weaning time is a great time to sit down and get this paperwork done.

Part of filling out the pedigree is assigning a tattoo number to each rabbit (see the sidebar "Show Rabbits Sport Tattoos" on page 25 in chapter 2). This number will be permanently tattooed inside each animal's left ear. Weaning day is also the perfect time to tattoo the new litter. For pet and meat rabbits, you may decide to skip tattooing as it's unnecessary.

Separate kits soon after they're weaned to avoid any conflicts or unintentional breeding.

The number 5 in this rabbit's ear (written in permanent marker) will be replaced by a real tattoo once the rabbit is assigned a number.

Now for culling. What exactly is *culling* and how is it done? Somewhere along the line, people got the impression (or were outright told) that culling means killing those rabbits that you don't want to keep in your rabbitry. While I suppose this might be an option for some, it is not what is meant by culling your herd. Culling is selecting which animals will be kept in your breeding program, which will be kept for show, which may be sold, and which will be used for meat. Culling is just about deciding where a rabbit is best suited, as every rabbit is born with different qualities from their siblings.

Proper culling is how your rabbitry herd will become a strong representation of the breed, fiber, or meat standard. In essence, during the culling process, you'll literally judge each rabbit for its intended purpose. If you breed French Lops for the show table, you may look at a young Frenchie to make sure it's healthy and has no inherent disqualifications, such as a malocclusion. You also want to compare it to the French Lop standard (as well as to your other show French Lops) and see how it holds up. A beautiful French Lop may not have the best ear placement for a winning show rabbit, but it could do well in youth shows or as a pet, and you may want to cull it from the rabbitry for either of those purposes.

It's wise to have a cage or several cages in an area of the rabbitry that's meant to temporarily house the rabbits you won't be keeping. This makes it easy for you to see at a glance what you have available when a phone call comes in from someone looking to purchase. Before you let any rabbit leave the premises, double-check that it hasn't changed over time and grown into a rabbit that you'd like to keep. You wouldn't be the first breeder to let go of a butt-kicking show rabbit because it was slow to mature and come into its own. In fact, most experienced rabbit-raisers will tell you that it's very difficult to decide which rabbits are worth keeping and which aren't when they're eight weeks old. Most people wait a couple more months to decide.

ADVICE FROM THE FARM

Bringing Up Baby

From making baby bunnies to caring for them, our experts have done it all.

Breeding Pointers

"The most important thing one must realize when breeding Holland Lops is that culling is a must. Hollands do not always breed true. You can breed two grand champions and end up with all pet bunnies. You can also breed two average Hollands and end up with phenomenal stock.

In breeding, try to always breed faults to good points. If one Holland has weak hips, do not breed it to another with weak hips, as that is all you will produce. Try to match up any faults in one rabbit to the good points in the one that you are breeding it to. This is not as easy as it sounds. You will often find yourself doing a balancing act when trying to match up faults with strong points. For example, a buck with a good crown and fair body could be bred to a doe with poor ear carriage and a good body. Ideally, you should always attempt to breed rabbits with the fewest faults."

—Chris Zemny, ARBA Judge

Earn Your Reputation

"Reputable breeders keep records on their animals, make responsible breeding choices, and have an effective plan to deal with the results. I also think it is important to be ethical when selling animals. If an animal isn't something you would be proud to have represent you on the show table, it isn't ethical to sell it to someone looking for a show rabbit or to start their own breeding program."

—Keelyn Hanlon, ARBA Judge

For First-Time Moms

"Some first-time does may not birth in the box, so it's a good idea to use a sheet of wood to cover almost half of the cage floor. If the doe does give birth outside of the box, gently move the kits to the box as soon as you notice them."

—Brenda Haas

chapter **SEVEN**

Making Money with Rabbits

Small-livestock keepers have long since figured out that rabbits can easily be kept for both pleasure and profit. While I don't think anyone claims that they're rolling in the riches thanks to rabbit keeping, a variety of rabbit products and by-products can certainly help offset your hobby costs as well as some household bills. Rabbit fanciers can sell their livestock as show animals, breeding stock, pets, and for meat. Consider also the fiber, manure, and fishing-worm markets. You'll definitely be able to strike a balance between your monthly budget and the creatures you love to keep.

Show Rabbits

Here's a bit of trivia: 800,000 rabbits are shown at rabbit shows in the United States each year. So yes, there's a market for selling show rabbits, and there are rabbit fanciers out there that show the breed you're raising. To acquire a good reputation for breeding high-quality rabbits, you'll first need to purchase some quality rabbits of your own. Then you'll need to begin breeding rabbits that are excellent representatives of the breed—so excellent, in fact, that they end up in the highest position possible on the show table. Because of this, marketing and selling show stock may take a little more time and money than some of the other options for rabbit-raising profits. Still, in my experience, showing rabbits and breeding for the fancy is the most rewarding and certainly the most fun opportunity.

Rabbit shows are inexpensive and easy to enter. Just hanging around at shows seems to get the fever going in just about anyone who's ever considered owning a rabbit. There are opportunities to make great friendships and to take home a wealth of information. In addition, many people make rabbit shows a family affair, with each family member helping to groom the rabbits or bring them to their show table. (For more information on showing your rabbits, see "Raising Show Rabbits" on page 23 in chapter 2.)

What to Pack for Rabbits that You're Selling

- Any comment cards (cards from shows that have the judges' notes on them) that pertain to the rabbits you are selling
- Pedigrees
- Poster board (or dry erase board) and markers to advertise that you have rabbits for sale
- Small baggies filled with rabbit food as a transitional feed for whoever buys your rabbits

If you'd like to raise show rabbits for profit, contact a local or national club that's devoted to the breed(s) that you're interested in raising. This is the best way to find the most reputable breeders for your breed. When it comes to rabbit breeders, I do need to warn you: as in every other area of life, there are unscrupulous breeders who want your money, with little concern as to what quality of rabbit ends up in your hands. These people can spot a newbie as he drives into the parking lot. To protect yourself, know as much as you can about which rabbit is best for you (and your family) and what the standard is for that breed beforehand. Remember that anyone who insists that you buy at that show or on that day is most likely waving a red flag in front of you. That isn't to say

If you're selling rabbits for exhibition, they should be pedigreed and of show-winning quality.

In addition to being show-quality, the rabbits you're selling should be accustomed to people and handling so that they can put their best foot forward at a crowded show.

that a reputable breeder can't have what you're looking for at the show. But most will listen to what you're looking for and not have any problem telling you to come by and see what they have at their home at a different time. Trust your gut.

As a show-rabbit-raiser, your goal should be, first and foremost, to breed rabbits of superior quality to what you have now in your rabbitry and to be a reputable ambassador for your rabbit breed. The show-rabbit world is smaller than you can even imagine. Rabbitries that rise to the top (and prosper) are those that are completely fair, honest, and helpful.

At this point, I'd like to impress upon you the responsibility that you have to your fellow breeders, to the rabbit owners who will come after you, and mostly, to the rabbits themselves to end the genetic line of certain negative inherited traits if they arise. I'm not talking about mismatched toenails or ears that are too short (although your goal would be to better that situation, as well). I'm speaking of issues such as malocclusion, entropion-affected eyelids (an eyelid that's folded under), and other physically disabling traits that make the life of the rabbit that has it uncomfortable or unlivable. Malocclusion can also be caused by a rabbit continually pulling on its cage wire, and entropion can be caused by a severe bacterial infection. However, both conditions are often the result of bad breeding. Nobody—especially the rabbits born in these litters—wins in this situation.

On the flip side, don't ever sell a rabbit that you know has faults or is ill to an unknowing buyer. Don't ever falsify papers or lie about an animal's age or how it's done on the show table. And don't even consider taking advantage of a 4-H kid. Your name will spread like wildfire through the show circuit in your area and beyond. A reputation of deceit or questionable ethics is hard to shake.

This English Angora rabbit is about to be a few pounds lighter thanks to wool harvesting.

Advertising is one of the most important aspects of any business. Most people end up selling their show stock through word of mouth (which is why it's that much more important to avoid any negative feedback). Breeders have cards printed up to give out at rabbit shows and white boards advertising what they have available. They are always watching for who's winning at the show table and will approach the breeders of winners to buy rabbits that will improve their own stock. Make yourself known to local 4-H groups, and advertise on bulletin boards at tack stores. And don't underestimate the power of the Internet! Create your own website, complete with pictures of your stock. See if you can get a listing or link on other breeders' websites as well as on club sites. If a show secretary will share his or her list of people who were sent show catalogs, you can always advertise using direct mail or e-mail. Take the opportunity to let show entrants know what rabbits you have available, and maybe offer a show special. Anything you can do to get your name out there and in a positive light will come back to you in sales.

Rabbit Fiber (Wool)

The first thing you should know about raising rabbits for fiber is that you won't get rich doing it. (For that matter, you probably won't get rich in the rabbit industry, period.) Still, if you're a spinner or a knitter and are involved in the fiber world in any way, you may find a nice little market for the by-product that your pet or show rabbits have to offer. Just as with other fiber, such as that of sheep or alpacas, rabbit wool can be color-dyed and is considered top drawer in the fiber-production market. Plush and silky Angora rabbit wool is one of the softest fibers in the world. The calm, docile Angora makes a wonderful pet, and if you're going to have one sitting on your lap, you might as well collect or spin its wool!

Angora rabbit wool comes one of two ways: raw or spun. Raw fiber is shorn or plucked from the rabbit (while the rabbit is molting) and sold for eight to sixteen dollars per ounce. Prices will vary depending on the quality of the wool, the length, and the percentage of guard hair. The softest part of the rabbit's coat is the undercoat, but there are longer, rougher hairs on the outside called guard hairs. The fiber that will bring in the most money will have high density (thickness), good crimp (waviness), and minimal guard hairs. If the rabbit has won awards for its coat, that can raise the price, too. Plucked wool may bring a higher profit than shorn fiber, but many Angoras today are being bred to hang on to that long fiber, making it necessary to shear it.

In addition to harvesting and selling raw fiber, you can spin your wool into skeins that are ready for knitters to use. Yarn and craft shops and knitting clubs alike will be interested in local, humanely harvested Angora wool. You can also dye rabbit wool to add color variety to the product. If you are a knitter, you have yet another opportunity for profit: you can create garments from the Angora wool you've harvested and spun from your own rabbits. Hats, scarves, mittens, and sweaters are all handmade items that enjoy timeless popularity at craft shows, especially just before the holiday season.

Many Angora breeders bring their raw or spun wool and knitted products to the rabbit shows to market. Fiber festivals, farmer's markets, and specialty shows are other possible markets for rabbit wool. And again, don't overlook the Internet as an instant marketing opportunity. Angora rabbit-raisers often market their raw fiber, spun skeins, and knitted products on their own websites or in online "stores" such as www.Etsy.com or www.MadeItMyself.com. Becoming involved with the National Angora Rabbit Breeders Club will get you well on your way to the wild and wooly world of rabbit fiber.

Derby's enjoying a little fuzzy companionship with to his oddly colored Angora friend.

Raise a Rare Breed

Today, the American Rabbit Breeders Association recognizes nearly fifty rabbit breeds (and counting). But there are some breeds whose numbers are dwindling; some of them rapidly so. These rabbits need help staying in existence. Some rabbit enthusiasts enjoy raising the rare breeds for conservation purposes as well as for showing or meat. It's personally satisfying to know that your hobby is making a difference in a breed.

This list is comprised from the American Livestock Breeds Conservancy (www.albc-usa.org), a nonprofit membership organization working to protect over 150 breeds of livestock and poultry from extinction. Endangered breeds are categorized into three lists that are labeled "Critical," "Threatened," and "Watch." Here are the ALBC's definitions for those categories and the rabbit breeds that are currently in them:

Critical
Fewer than 200 annual registrations in the United States and an estimated global population of fewer than 2,000.

- American
- American Chinchilla
- Silver Fox

Threatened
Fewer than 1,000 annual registrations in the United States and an estimated global population of fewer than 5,000.

- Belgian Hare
- Blanc de Hotot
- Silver

Watch
Fewer than 2,500 annual registrations in the United States and an estimated global population of fewer than 10,000. This list also includes breeds that present genetic or numerical concerns or that have limited geographical distribution.

- Beveren
- Creme d'Argent
- Giant Chinchilla
- Lilac
- Rhinelander

Rabbit Meat

While European countries such as France, Spain, Italy, and Germany have long been utilizing this lean and nutritious meat, rabbit has barely begun to touch the meat industry in America. Meat rabbits are easy to raise, and the market for them has huge growth potential here in the United States. There's a nice little niche for this type of market, beginning with friends and family and eventually expanding to reach coworkers, local restaurants, and small grocery stores. As I mentioned in chapter 2, certain breeds are considered the most profitable when it comes to raising rabbits for meat. The New Zealand and Californian are at the top of the list, but you should decide which breed is right for you as you investigate this potential market.

The most profitable way to sell rabbit meat is if you do the processing (butchering) and dressing (cleaning and packaging) yourself. However, doing so creates a lot of work and requires more than just basic knowledge. You must also know and comply with your local health department's requirements. Before processing rabbits at home, you first need to explore and understand those regulations, as well as any city zoning laws that might be in place. On the plus side, you don't need any special licensing as long as you're following those laws.

Even if there aren't any zoning restrictions in your area, you still need to take your neighbors into consideration. For one thing, they may not be pleased if you do your processing where their children can watch. But if your neighbors are on board with your hobby-farming practices, they may constitute a whole new network of buyers. Being considerate of those around you is both practical and kind.

Not everyone can stomach processing rabbit meat. Many meat-rabbit breeders opt to send their fryers (young meat animals) to an outside processor instead of doing the deed themselves. The processor will then return the processed meat for you to market and sell. If you can find a source to sell your live rabbits to directly who will process and sell the meat themselves, you can cut out the cost of the middleman. Selling live rabbits to a processor brings in the least amount of money, but it's also the easiest way to run your business with very little effort on your part. If you'd like to find a rabbit processor near you, check out the list on the American Rabbit Breeders Association at www.arba.net/Processors.htm.

Researching your local regulations and planning your business approach *before* you begin breeding rabbits for meat are very important. Rabbits are usually sold as fryers when they reach 3½ pounds or more, which is around ten to twelve weeks old—that's not a lot of time to do all of the footwork involved in complying with laws and getting the word out.

Just as important as making sure you're abiding by all laws, you need to figure out if you have a market for rabbit meat in your area—again, preferably *before* deciding to start a rabbitry for meat rabbits. Otherwise, you may end up looking for pet homes for a lot of rabbits. You can go about studying the possible consumer demographics in your area the very same way that you would if you already had the rabbitry up and running—advertise your product at local farmer's markets, feed stores, newspapers, local classified pages, and by e-mail if you have a healthy list of rabbit or organic-consumer contacts. Don't forget to approach local restaurants and specialty grocery stores.

About the Pet Rabbit Market

Whether or not it's ethical to get into the pet rabbit market is debatable among rabbit breeders. If you breed and sell pets responsibly (in small numbers), this may be a good outlet for you. The most frequent argument against selling pet rabbits is the same as the one against breeding cats or dogs—many unwanted rabbits wait for homes in animal shelters everywhere. This should certainly be taken into consideration before jumping into any pet-selling industry. But it's certainly worth discussing your options so that you can make up your own mind.

Let's talk about culling, which forces us to look at selling pet rabbits from a practical and humane standpoint. When a rabbitry has culled those rabbits that are deemed unfit as show or meat animals, the fate for the rabbits that didn't make the cut is usually euthanasia. Obviously, selling the rabbit as a pet instead is a humane alternative. In these cases, a rabbit breeder will typically find his or her pet-quality rabbits homes through word of mouth rather than sell them to a pet store. The breeder might ask friends and neighbors to spread the word, advertise in classifieds, or even turn up at local shows to try to reach attendees who are interested in rabbits for themselves and not just for showing.

Some people will breed extra litters specifically to sell as pets to the general public. However, in an attempt to retain some control over who purchases their animals, as well as to educate potential pet owners, these breeders sell pets directly from their rabbitries as opposed to through a pet store where they have zero control as to where they end up. Usually, breeders will also offer lifetime rabbit-care support to those who take home their rabbits.

Rabbits have unique personalities and can make lovable pets for all ages.

Selling rabbits to pet stores for the sole purpose of creating more pets (and quick cash) is the rabbit industry that's most frowned upon. It isn't that the breeder in this case is a "bad" breeder by definition, but rather that the rabbits are treated solely as a product—not as a living creature—by the pet shop industry. Therefore, less concern is often placed on their welfare once they're inside the pet store and out of the breeder's hands. As a responsible rabbit breeder, your first concern should be where your offspring will end up and how well they will be cared for.

Rabbit Manure

Rabbit manure is a natural by-product of your rabbitry and is precious as far as gardeners are concerned. Seasoned gardeners will tell you that rabbit manure is, hands down, the very best manure for creating magnificent compost. It's the most nutritionally balanced of any of the herbivore manures, and it produces extensive positive effects on the garden bed. As the manure breaks down, it builds soil structure, improves soil porosity, augments soil stability, and holds nutrients not only for plants but also for other soil organisms.

As a rabbit keeper with manure to spare (for a price), you should tap into this market by advertising online, placing fliers on bulletin boards (nurseries are the perfect place), and contacting garden clubs and community gardens. Spreading the word locally is key because you won't want to ship this particular product. There's no "going rate" for rabbit manure. But if you can fill a garbage bag or burlap sack while cleaning cages (shavings and bedding mixed in are fine), you can ask for four or five dollars. (This price may vary depending on where you live.) (See "Rabbit Manure as Garden Gold" on page 36 in chapter 2 for more information on composting.) That doesn't sound like much, but that depends on how many rabbits you have in your rabbitry! If you don't have a compost pile (although I encourage you to start one) and aren't raising worms, then you may just be thankful to have the manure hauled away for free by zealous gardeners.

The Rabbit–Red Worm Connection

As I mentioned in chapter 2, if there were ever a perfect companion to rabbits, red worms are it. While all worms can be used for composting, the ones with the voracious appetite for rabbit manure are Red Wigglers (*Eisenia fetida*) or Red Tigers (*Eisenia andrei*). These are surface-dwelling worms that munch on the debris left on the ground, quickly turning it into nutritional castings for gardens. When worms are raised in rabbit manure, the castings are mixed in with the composted material from the manure (see the sidebar "DIY Vermicomposting" on page 38 in chapter 2 for more information). Just to be clear, this actually creates vermicompost (compost that has worm-composted organic materials in the mix) as opposed to pure worm castings (without other composted materials). Vermicompost is exactly what worm farms produce from kitchen scraps.

Myth Buster: The Rabbit Died

"The rabbit died" has been whispered behind doors or shouted from the rooftops both in real life and in Hollywood movies, letting the world know a woman was pregnant. In 1927, scientists discovered that pregnant women produce a hormone called hCG (Human Chorionic Gonadotropin). They found that when a pregnant woman's urine is injected into a rabbit and hCG was present, the rabbit's ovaries begin to change. The popular phrase suggests, misleadingly, that the rabbit dies *because* the woman is pregnant. But, of course, the rabbit only died because the doctor had to observe its ovaries.

Modern pregnancy tests still measure hCG levels—but now, luckily for rabbits everywhere, with no animal involvement. But old habits die hard, and some people still announce a pregnancy with "The rabbit died!"

This custom hutch has a wooden bin underneath it to catch droppings. The bin also contains worms for vermicomposting so that unwanted manure can fertilize lush gardens.

You can use the vermicompost that's created under your rabbit cages in your own garden, but you can also sell the extra red worms to fishermen. The investment for this type of business is minimal—small containers for the worms and maybe some marketing efforts. As far as containers go, just be sure that you are able to poke air holes in whichever ones you choose. And don't forget to label the containers with your name and number should someone want to purchase more worms.

If you'd like to stand out in a crowd of fishing-worm raisers, you can place worms into an interim container and fatten them up using a special supplemental diet (as opposed to just the manure that they feast on otherwise) a couple of weeks before putting them on the market. One such diet consists of 5 parts chicken layer mash (from your local feed store), 2 parts wheat or rice bran, 1 part agricultural lime, 1 part wheat flour, and 1 part powdered milk. There are many more worm-fattening diets out there you can try—just ask a seasoned fisherman!

Market your worms by contacting local fishing or sporting-good and convenience stores that are located near popular fishing spots. Be sure to let fishing clubs, summer camps, and outdoor youth organizations such as the Boy and Girl Scouts of America or 4-H know that you're in business. Visit nearby marinas and piers and talk to fishermen or hang fliers (first get permission to do so). Be aware that selling worms for fishing may only be a seasonal business depending on where you live.

ADVICE FROM THE FARM

Making and Saving Money with Rabbits

Our rabbit-raising experts have some thoughts on money and the rabbitry.

Two for the Show

"If you are serious about breeding Holland Lops, the best suggestion I have for you is to purchase the best buck that you can afford. You will need to get a picture in your head of what a perfect Holland Lop looks like, and find a buck that comes as close to your picture as possible. If the perfect buck comes with a high price tag, wait, save, and buy the rabbit. The investment will be worth it. Just remember that when you are linebreeding, you always return to your best animal, which is usually your herd buck. He is the most important acquisition to your breeding program. You will use him generation after generation."

—Chris Zemny, ARBA Judge

A Penny Saved Is a Penny Earned

"When buying your cages, I would recommend going to ARBA shows and buying cages, carrier cages, and other supplies there. Most shows have vendors who make cages and sell them at a better price than you would find at pet stores. You can make some profit by getting your rabbitry's name out by making business cards or attending plenty of shows to sell your stock at."

—Sarah Haas

Diversify, Diversify, Diversify

"I raise some rabbits for meat that I can sell to local restaurants; this makes my rabbit project a profitable one."

—Patrick J. Hinde III

Acknowledgments

When I discovered rabbits two decades ago, I didn't fall in love with *one* rabbit, I fell in love with the rabbit world as a whole. The book you're holding in your hands is the rabbit book I've wanted to write for many years. I wrote it for rabbit lovers that want to take that exciting step beyond enjoying a single pet rabbit to becoming the keeper of an entire rabbitry and raising a small herd. It's my hope that you'll enjoy reading this book as much as I enjoyed writing it.

A big thank you to Andrew DePrisco, Karen Julian, and Jennifer Calvert at BowTie Press for being so wonderful and generous with their time and encouragement. I want to send monstrous thank yous to all the gracious people I chased down to get quotes for the "Advice from the Farm" sidebars (because everyone was at the ARBA convention!). You are my people.

I have to give some extra shout-outs to Jennifer Ambrosino and Linda Hoover as I probably bent their ears (and their emails) the most while getting this book done.

I can't forget to thank my kids who pretty much lived with a crazy-mother-in-hiding while a book was being born. A special thank you to my husband-extraordinaire, Bobby, who has spent countless hours at rabbit shows, building shelters, and warming Popsicle bunnies. You'll never know how appreciated you are for your patience and support of this writer and animal lover.

Glossary

abscess—a localized collection of pus caused by an infection.

altricial—born blind, naked and helpless.

agouti—a rabbit coat pattern that is salt-and-peppered. Wild rabbits have an agouti patterned coat.

Angora—a rabbit that produces wool instead of regular fur.

breed—a group of rabbits that share the same characteristics such as body type, coat, size, etc.

buck—a male rabbit.

coccidiosis—this aggressive condition caused by an overabundance of parasites can be lethal for rabbits.

compactness—a body style that's desirable in certain rabbit breeds. A compact body would be short and stocky.

conditioning—grooming, feeding, and other top quality preparations to bring a rabbit to its full potential. Good condition would include firm flesh, a shiny coat, and clear, bright eyes.

coprophagy or cecotrophy—the eating of feces, which is a normal nutritional requirement for rabbits.

crepuscular—describes animals such as rabbits that are most active at twilight, dawn, or dusk.

crossbreed—an animal that's born from parents of two different breeds.

cull—culling refers to separating rabbits according to how one intends to use them. It can mean to actually eliminate the culls from the herd entirely.

Cuterebra fly—a flying insect that lays its eggs on rabbits. The larva lives as a parasite under the rabbit's skin.

density—the thickness of a rabbit's coat.

dewclaw—the extra claw on the side of the front feet. This claw lacks a bone like the other four toes.

dewlap—a fold of loose skin under the chin of some female rabbits. It's considered normal but could be a disqualification on the show table for some rabbit breeds.

disqualification—a defect that eliminates a rabbit from competing either temporarily or permanently in shows.

doe—a female rabbit.

domesticated—adapted to being around (and/or trained) by humans.

dressed—a rabbit that's been skinned and prepared for cooking.

dress-out—the percentage of edible meat available once dressed.

dwarf—a rabbit that weighs no more than 3 pounds at maturity.

ear canker—scabby condition inside the rabbit's ear that's caused by ear mites.

fostering—giving a nursing doe a kit to raise that isn't from her own litter.

genitals—the rabbit's reproductive organs, usually referring to those on the outside of the body.

gestation—the period of time from mating to the birth (usually thirty-one days).

guard hair—the longer, coarser hair on a rabbit that's mixed in with the softer coat.

hare—a mammal that's also in the order Lagomorpha along with rabbits but has larger feet, longer ears, and different habits.

heat stroke—a condition that comes from being exposed to high temperatures and can cause death in a rabbit.

hock—the first joint on the hind leg of a rabbit, usually thick furred.

hutch—a rabbit house. Either what was traditionally called a *hutch* such as a homemade wood and wire enclosure or any rabbit cage.

hutch card—a record-keeping card that's attached to the rabbit cage which has pertinent information on each rabbit, such as birth date or breeding date.

inbreeding—breeding close rabbit relatives. Can be advantageous when done correctly.

inherited—a characteristic that's passed down from a parent or other ancestor to the next generation.

instinct—an animal's natural tendency to behave in a certain way.

junior—a rabbit that's under six months old.

kindle—giving birth to a litter of kits.

kit—a baby rabbit; short for kitten.

lagomorph—a mammal that's in the order Lagomorpha; which are rabbits, hares, and pikas.

linebreeding—a system of breeding rabbits so that offspring are close descendants. This creates a "line" of rabbits.

litter—kits or other baby animals born at the same time from the same mother.

malocclusion—a dental defect that doesn't allow the mouth to close properly.

mite—a very tiny parasite that can cause the scabby condition called an "ear canker" inside a rabbit's ear.

molting—the condition while shedding fur.

mucoid enteritis—this disease is usually seen in young rabbits. Symptoms include jelly-like diarrhea, increased thirst, and loss of appetite. Often fatal.

nest box—a box where the doe will give birth to her kits (or kindle).

outcrossing—breeding two animals of the same breed that are unrelated to each other.

palpate—within the context of this book, to palpate is to feel for young through the doe's abdomen with your hands.

peanut—those tiny kits in a litter that lack the proper growth tissue to survive and are usually half the size of their siblings. They're often found in litters of the breeds that carry the dwarf gene. They're not the same thing as the "runt of the litter." Peanuts don't survive long.

pedigree—a record of a rabbit's ancestry, ideally including at least three generations recorded on it; four or five generations are optimal.

precocial—those animal young that are relatively mature and mobile from the moment of birth (or hatching). Hares are precocial as are chickens.

preventative—measure taken to ward off problems or disease before the fact.

purebred—an animal of a recognized breed from parents and other ancestors of that breed.

rabbitry—the place where the rabbits are kept and raised. An area where all of the hutches are situated.

registered—different from a pedigree, listed with an organization. A rabbit can only be registered by an ARBA official. If it passes, the registrar records it and puts a special tattoo in the rabbit's right ear.

rodent—a mammal belonging to the order Rodentia, including mice, rats, and squirrels.

senior—in a rabbit breed that weighs under 10 pounds at maturity, a "senior" identifies a rabbit over six months old. In a rabbit breed that matures to over 10 pounds, senior would describe an eight-month-old.

snuffles—a highly contagious respiratory disease. One of the symptoms is nasal discharge.

sore hocks—this condition describes ulcerated footpads on a rabbit.

tattoo—permanent ear markings that are used to identify a show rabbit. Rabbitry (the breeder's) tattoos are placed in the left ear and registration tattoos in the right.

territorial—describes a rabbit that's protective over its living quarters (hutch).

type—the body conformation of a rabbit.

weaning—the time when the young rabbits are separated from the doe (about six to eight weeks old).

wool block—a blockage in the digestive tract that's caused by the rabbit's ingesting wool or fur.

Resources

Clubs and Associations

American Belgian Hare Club
Al Butler
330-484-1977
www.belgianhareclub.com

American Beveren Rabbit Club
Meg Whitehouse
732-919-0909
www.freewebs.com

American Britannia Petite Rabbit Society
Ron Rohrig
765-966-4226
www.britanniapetites.com

American Checkered Giant Rabbit Club, Inc.
David Freeman
513-576-0804
www.acgrc.com

American Chinchilla Rabbit Breeders Association
Diana Young
812-339-0040

American Dutch Rabbit Club
Sue Hill
608-832-1579
www.dutchrabbit.com

American Dwarf Hotot Rabbit Club
Tonna Thomas
417-462-7146
http://www.adhrc.com

American English Spot Rabbit Club
Michael C Wiley, Sr.
502-535-7051
http://americanenglishspot.webs.com/

American Federation of New Zealand Rabbit Breeders
John Neff
407-349-0450
www.newzealandrabbitclub.net

American Fuzzy Lop Rabbit Club
Muriel Keyes
503-254-2902
http://users.connections.net/fuzzylop

American Harlequin Rabbit Club
Pamela Granderson
812-843-5460
www.americanharlequinrabbitclub.net

American Himalayan Rabbit Association
Errean Kratochvil
727-847-1001
www.himalayanrabbit.com

American Netherland Dwarf Rabbit Club
Susan Clarke Smith
828-586-9698
www.andrc.com

American Polish Rabbit Club
Patti Walthrop
817-312-0305
www.americanpolishrabbitclub.com

American Rabbit Breeders Association
309-664-7500
www.arba.net

American Sable Rabbit Society
Dennis E. Frost
316-944-0522
www.americansables.webs.com

American Satin Rabbit Breeders Association
Rita Peralta
209-931-0983
www.asrba.org

American Standard Chinchilla Rabbit Breeders Association
Patricia Gest
941-729-1184
www.ascrba.com

American Tan Rabbit Club
Virginia Akin
325-236-4032
www.atrsc.org

American Thriantha Rabbit Breeders Association
Carrol G. Hooks
254-986-2331
http://atrba.net

Breeders of the American Rabbit N.S.C.
Chris Hemp
831-227-3939
www.americanrabbits.org

California Rabbit Specialty Club
Susan Yeary
956-383-2228
www.nationalcalclub.com

Champagne d'Argent Rabbit Federation
Lenore Gergen
651-283-0202
E-mail: mccavy@aol.com

Cinnamon Rabbit Breeders Association
Stacie Snider
972-768-2916
www.cinnamonrba.webs.com

Creme d'Argent Rabbit Federation
Travis W. West,
740-698-3014
E-mail: cremedargent@hotmail.com

Florida White Rabbit Breeders Association
Jane Meyer
217-387-2427
www.fwrba.net

Giant Chinchilla Rabbit Association
Mary Ellen Stamets
908-782-0462
http://giantchinchillarabbits.webs.com

Havana Rabbit Breeders Association
Suzanne Marie Hemsath
801-250-9836
http://havanarb.org

Holland Lop Rabbit Specialty Club
Pandora M. Allen
757-421-9607
www.hlrsc.com

Lop Rabbit Club of America
Sandy Bennett
803-755-3122
www.lrca.us

Mini Lop Rabbit Club of America
Pennie Grotheer
417-842-3317
http://minilop.org

National Angora Rabbit Breeders Club
Margaret Bartold
573-384-5866
www.nationalangorarabbitbreeders.com

National Federation of Flemish Giant Rabbit Breeders
Karen S. Clouse
260-894-1202
www.nffgrb.com

National Jersey Wooly Rabbit Club
Amanda Pitsch
616-498-2330
www.njwrc.net

National Lilac Rabbit Club of America
Judy Bustle
704-624-5465
http://nlrca.webs.com

National Mini Rex Rabbit Club
Jennifer Whaley
619-933-6505
www.nmrrc.net

National Rex Rabbit Club
Arlyse Deloyola
512-392-6033
http://rexrabbit.tripod.com

National Silver Fox Rabbit Club
Lee Ann Schneegas-Nevills
607-404-7501
www.nsfrc.com

National Silver Rabbit Club
Laura Atkins
816-732-6208
www.silverrabbitclub.com

North American Lionhead Rabbit Club
Jennifer Hack
765-346-7604
www.lionhead.us

Palomino Rabbit Co-Breeders Association
Deb Morrison
918-396-3587
www.palominorabbit.com

Rhinelander Rabbit Club of America
Wesley Planthaber
814-667-2406
www.rhinelanderrabbits.com/main.htm

Silver Marten Rabbit Club
Stephanie Myers
541-289-2444
www.silvermarten.com

Rabbit Supplies

American Rabbit Breeders Association Links
www.arba.net/Links.htm
This is ARBA's link page that has rabbit-supply companies belonging to ARBA members.

Bass Equipment Company
www.bassequipment.com
Cages, waterers, and feeders, oh my! Everything you need to set up your small-scale rabbitry.

Bunny Rabbit
www.bunnyrabbit.com
This is a fun sight geared toward the rabbit enthusiast that wants to bring his or her bunnies into their daily lives with jewelry, home decor, purses, and T-shirts. Some of the items are provide by an affiliate but many are handmade and exclusive to Bunny Rabbit.

Jeffers Livestock
www.jefferslivestock.com
These guys have been around a while (1975) and many small farms like their products and their prices.

Klubertanz Equipment Company
www.klubertanz.com
Weld-wire supplier, cage building supplies, medications, etc.

KW Cages
clover.forest.net/kwcages/index.html
More wonderful rabbit goodies. This company has some glorious stackables.

Rabbit Mart
www.rabbitmart.com
This site has more for the pet rabbit set-up. But they have decent prices and some interesting cages.

Valley Vet Supply
www.valleyvet.com
This is a general animal- and livestock-supply company. It's extremely popular as an online source for small farms of all types.

Websites

American Livestock Breeds Conservancy
http://albc-usa.org
The American Livestock Breeds Conservancy is the pioneer organization in the United States working to conserve heritage breeds and genetic diversity in livestock.

Association of Exotic Mammal Veterinarians
www.aemv.org
This website can help you find an exotic animal vet (which is the classification for rabbits) in your area.

Bjerner's Rabbit Hopping Site (UK)
www.kaninhop.dk/uk
Here's where you want to be, certainly if you're in the United Kingdom and even if you're in the United States. Super information here, courtesy of rabbit-hopping expert, Aase Bjerner.

Rabbit Hopping USA (Website)
www.rabbithopping-usa.webs.com
If you're interested in the sport of rabbit hopping in the United States, here's the place to start.

Rabbit Hopping USA (Yahoo Group)
http://pets.groups.yahoo.com/group/RabbitHoppingUSA-Agility
This is a very active rabbit-hopping Yahoo group run by the extremely helpful and knowledgeable Linda Hoover.

Rabbit Web
www.rabbitweb.net
This is a general information website that has a rabbit forum attached.

Raising Rabbits in the Pacific Northwest
http://pan-am.uniserve.com/
This is a resource for the commercial meat rabbit breeder.

Rudolph's Rabbit Ranch
www.rudolphsrabbitranch.com/rrr.htm
This is a pretty thorough site for information on raising meat rabbits.

The Independent Pet and Animal Transportation Association International
903-769-2267
www.ipata.com
These guys will help you get your rabbits from here to there.

The Joy of Handspinning
www.joyofhandspinning.com/angora-care.shtml
This is a terrific website that will steer you in the right direction if you'd like to spin your Angora fiber.

National 4-H Club
www.4-h.org
4-H is one of the most valuable youth organizations anywhere in the world. If you have children, please come visit. If you have children and rabbits, this is definitely the place for you.

National Association of Professional Pet Sitters
856-439-0324
www.petsitters.org
Leaving town without the rabbits? Find yourself a professional pet sitter in your area who's familiar with caring for rabbits and insured.

Pets Welcome
www.petswelcome.com
If you find yourself traveling with your rabbits, this is the place to find pet-friendly hotels.

Pet Sitters International
336-983-9222
www.petsit.com
The best place to find top-quality pet sitters in your area that know the ins and outs of rabbit care.

Rabbits Online
www.rabbitsonline.net
A rabbit lover's forum.

USDA Animal and Plant Health Inspection Service
www.aphis.usda.gov/animal_welfare/pet_travel/pet_travel.shtml
Shipping or traveling with rabbits? This is the website that'll give the info you need to know.

More Great Books on Rabbit-Raising

American Rabbit Breeders Association, Inc. *The Official Guide Book to Raising Better Rabbits & Cavies.* 2000.
This publication is a compendium of sections and articles written by some of the most seasoned and well-known rabbit-raisers in America.

Bennett, Bob. *Storey's Guide to Raising Rabbits.* Storey Publishing, 2001, 2009.
Bob Bennett is one of the most revered rabbit breeders in our time. His sound advice has been sought after for many years. His book is well detailed.

Damerow, Gail. *Barnyard in Your Backyard: A Beginner's Guide to Raising Chickens, Ducks, Geese, Rabbits, Goats, Sheep, and Cattle.* Storey Publishing, 2002.
This book contains concise, but excellent rabbit-raising advice for those who are living with animals on small acreage.

Mays, Howard L. *Raising Fishworms with Rabbits.* Shields, 1981.
This is a slender book, but it gives the ins and outs of making a profit from fishing worms raised under rabbit cages.

McLaughlin, Chris. *The Complete Idiot's Guide to Composting.* Alpha, 2010.
If you want to make the best compost, be sure to use your rabbit manure in your compost piles. This book tells you everything about composting with rabbit manure and everything else. It also has a whole section dedicated to vermicompost (worm farms).

Searle, Nancy. *Your Rabbit: A Kid's Guide to Raising and Showing.* Storey Publishing, 1992.
Nancy Searle hits the most important highlights in her book on rabbit-raising and showing. It's the handiest book around for kids getting into the show circuit, and adults will find it just as valuable.

Photo Credits

The sources for the photographs for this volume are listed below. Images ending in "/SS" are from Shutterstock.com. Images ending in "/Flickr" are from Flickr.com.

Front Matter
2 artemisphoto/SS
3 Diana Taliun/SS
6 CoolR/SS
8 misha shiyanov/SS

Chapter 1
10 Rick Wylie/SS
13 CoolR/SS
14 Matthijs Wetterauw/SS
16 Dudarev Mikhail/SS
17 Lucian Coman

Chapter 2
18 Nataliya Peregudova/SS
21 marylooo/SS
22 Heather LaVelle/SS
24 The Original Turtle/Flickr
25 Lauri Rantala/Flickr
28 Tomi Tapio/Flickr
29 Billy Hathorn/Wikimedia
30 Höstblomma/Wikimedia
31 Andreas Altenburger
32 Teresa Levite
35 Natalia Wilson/Flickr
36 Hgalina/SS
37 Kokhanchikov
39 Jessica Reeder/Wikimedia

Chapter 3
42 Francesco83/SS
45 Neil Bird/Wikimedia
46 Charlene Bayerle/SS
47 Regien Paassen/SS
49 Foshie/Flickr
50 Roberto Reyes/Wikimedia
51 Benny Mazur/Flickr
52 Richard Peterson/SS
53 (top) frantab/SS
(bottom) Oosoom/Wikimedia
56 Joy Brown/SS
58 2009fotofriends/SS
59 Petar Ivanov Ishmiriev/SS
61 Lisa Jacobs/Flickr
62 Regien Paassen/SS

Chapter 4
64 Juice Team/SS
66 wim claes/SS
67 GRISHASS
68 Kassia Halteman/SS
69 Carly Lesser & Art Drauglis/Flickr
70 Daniel Tay/SS
71 Carly Lesser & Art Drauglis/Flickr
72 Ostanina Ekaterina Vadimovna/SS
73 Carly Lesser & Art Drauglis/Flickr
74 Tomi Tapio/Flickr
75 The Original Turtle/Flickr
76 SBriggs

Chapter 5
78 Tyler Olson/SS
80 STILLFX/SS
82 (main) Heike Rau/SS
(inset) Aprilphoto/SS
85 (top) xjrshimada/SS
(bottom) Carly Lesser & Art Drauglis/Flickr
86 Carly Lesser & Art Drauglis/Flickr
87 spiraltri3e/Flickr
88 Heidi Brand/SS
89 Eric Isselée/SS
90 Carly Lesser & Art Drauglis/Flickr
91 Jannes Pockele/Flickr
92 Sue McDonald/SS
95 Grandpa/SS
96 Dmitriy Shironosov/SS
97 Katrina Brown/SS

Chapter 6
98 Nikki Gibson/Flickr
101 Justin + Elise Snow/Flickr
102 Robobobobo/Flickr
105 The Original Turtle/Flickr
106 The Original Turtle/Flickr
107 The Original Turtle/Flickr
109 (both) Nikki Gibson/Flickr
110 Nikki Gibson/Flickr
113 The Original Turtle/Flickr
114 Julija Sapic/SS
115 Feng Yu/SS
116 The Original Turtle/Flickr
117 Imageman/SS

Chapter 7
118 Carly Lesser & Art Drauglis/Flickr
120 Benny Mazur/Flickr
121 Tom Woodward/Flickr
122 Nick Bramhall/Flickr
123 Carly Lesser & Art Drauglis/Flickr
126 Amanda Warren/Flickr
128 Chris McLaughlin
129 Keattikorn/SS

Back Matter
130 TOMO/SS
132 ansem/SS
135 constructer/SS
136 Ostanina Ekaterina Vadimovna/SS
141 Caroline VancoillieSS
143 Knumina/SS
144 piyato/SS
146 Vera Kailova/SS
151 David Ryznar/SS
154 Cherkas/SS
159 ravl/SS

Index

D

E

F

Q

R

V

W

Z

about the **AUTHOR**

Chris McLaughlin is a freelance writer and author living in Northern California. She's raised rabbits for seventeen years, and many of those years were spent breeding and showing American Fuzzy Lops as well as leading rabbit projects for 4-H clubs. Chris also founded a rabbit rescue in the Sierra Foothills and worked at Sierra Wildlife Rescue in Placerville, California, where her specialty was rehabilitating and releasing wild rabbits.

Today, Chris is a contributor to *Urban Farm Magazine* and the author of several gardening books. She and her family share their hobby farm with rabbits and many other animal species in the California Gold Country.